Health Information Science

With the development of database systems and networking technologies, Hospital Information Management Systems (HIMS) and web-based clinical or medical systems (such as the Medical Director, a generic GP clinical system) are widely used in health and clinical practices. Healthcare and medical service are more data-intensive and evidence-based since electronic health records are now used to track individuals' and communities' health information. These highlights substantially motivate and advance the emergence and the progress of health informatics research and practice. Health Informatics continues to gain interest from both academia and health industries. The significant initiatives of using information, knowledge and communication technologies in health industries ensures patient safety, improve population health and facilitate the delivery of government healthcare services. Books in the series will reflect technology's cross-disciplinary research in IT and health/medical science to assist in disease diagnoses, treatment, prediction and monitoring through the modeling, design, development, visualization, integration and management of health related information. These technologies include information systems, web technologies, data mining, image processing, user interaction and interfaces, sensors and wireless networking, and are applicable to a wide range of health-related information such as medical data, biomedical data, bioinformatics data, and public health data.

More information about this series at http://www.springer.com/series/11944

Dong Xu · May D. Wang
Fengfeng Zhou · Yunpeng Cai
Editors

Health Informatics Data Analysis

Methods and Examples

Springer

Editors
Dong Xu
Digital Biology Laboratory, Computer Science Department
University of Missouri-Columbia
Columbia, MO
USA

May D. Wang
Georgia Institute of Technology and Emory University
Atlanta, GA
USA

Fengfeng Zhou
College of Computer Science and Technology
Jilin University
Changchun
China

Yunpeng Cai
Shenzhen Institutes of Advanced Technology
Chinese Academy of Sciences
Shenzhen, Guangdong
China

ISSN 2366-0988 ISSN 2366-0996 (electronic)
Health Information Science
ISBN 978-3-319-44979-1 ISBN 978-3-319-44981-4 (eBook)
DOI 10.1007/978-3-319-44981-4

Library of Congress Control Number: 2017941057

Printed on acid-free paper

This Springer imprint is published by Springer Nature
The registered company is Springer International Publishing AG
The registered company address is: Gewerbestrasse 11, 6330 Cham, Switzerland

Preface

In the past decade, we have witnessed tremendous growth in biomedical data generation and substantial improvement of computational capacities (both hardware and computational methods) that can be used to handle these data. As a result, these "Big Data" provide great opportunities to health informatics and healthcare in general. In particular, the available data and the data-driven approach have started to empower precision medicine, which provides personalized disease treatment and prevention by taking into account individual variability in genes, environment, and lifestyle. On the other hand, the huge amount of the data and how to use these data raise unparalleled challenges to data scientists and informatics researchers. It is highly nontrivial to provide useful computer-aided analyses of heterogeneous biomedical datasets accumulated in various databases and electronic health records (EHRs). The biomedical data are notorious for its diversified scales, dimensions and volumes, and require interdisciplinary technologies for visual illustration and digital characterization. Various computer programs and servers have been developed for these purposes. But how to choose and use them are often difficult, especially for beginners. In addition, integrating different data and tools together to assist medical diagnosis and treatment is even more challenging.

A number of edited books have been published to discuss different aspects of health informatics data analysis. However, these books typically focus more on individual research. The authors of each chapter often emphasize their own methods. There is lack of comprehensive overview for the field, and hence the existing books are often difficult for beginners. This book is an attempt to systematically review the computational methods and tools for different aspects of health informatics data analyses. We have designed this handbook to comprehensively cover major topics in the field, as well as to provide concrete examples. Each chapter provides the detailed review of the state-of-the-art computer programs and an example procedure of data analysis and data fusion for each of 13 important biomedical questions. By following the step-by-step procedure, you will be exploring the biomedical questions with various programs and servers like a pro. Each chapter in the book is a self-contained review of a specific topic. Hence, a reader does not need to read through the chapters sequentially. A brief description of each chapter is given below.

Chapter "ECG Annotation and Diagnosis Classification Techniques" reviews the general techniques of ECG beat annotation and classification. It shows a preliminary study on deep learning application in ECG classification, which leads to better results and has a high potential both for performance improvement and unsupervised learning applications.

Chapter "EEG Visualization and Analysis Techniques" presents the current status of EEG research with projected applications in the areas of health care. As an example, it describes a method of quick prototyping an EEG headset in a cost-effective way and with state-of-the-art technologies.

Chapter "Biomedical Imaging Informatics for Diagnostic Imaging Marker Selection" discusses challenges and techniques of biomedical imaging informatics in the context of imaging marker extraction. In particular, it focuses on how to regulate image quality, extract image features, select useful features, and validate them.

Chapter "Big Health Data Mining" demonstrates different data levels involved in health informatics and introduces some general data mining approaches. An example case study is illustrated for mining long-term EHR data in epidemiological studies.

Chapter "Computational Infrastructure for Telehealth" introduces telehealth systems and their computational architecture, as well as challenges associated with creation of the 'complete-loop' solution. It also includes a practical use case describing an application for monitoring patients with hypertension.

Chapter "Healthcare Data Mining, Association Rule Mining, and Applications" introduces popular data mining algorithms and their applications in health care. It focuses on association rule mining that can provide a more flexible solution for personalized and evidence-based clinical decision support.

Chapter "Computational Methods for Mass Spectrometry Imaging: Challenges, Progress, and Opportunities" examines current and emerging methods for analysis of mass spectrometry imaging (MSI) data. It highlights associated challenges and opportunities in computational research for MSI, especially in proteomics, lipidomics, and metabolomics with spatially resolved molecular information.

Chapter "Identification and Functional Annotation of lncRNAs in Human Disease" describes the current bioinformatics methods to identify long noncoding RNAs (lncRNAS) and annotate their functions in mammal. It also provides several ways to further analyze the interactions between lncRNAs and targets, such as miRNAs and protein coding genes.

Chapter "Metabolomics Characterization of Human Diseases" summarizes popular bioinformatics analysis tools for characterizing human diseases based on their metabolomics profiles. Pathway analysis using metabolite profiles and disease classification using metabolite biomarkers are presented as two examples.

Chapter "Metagenomics for Monitoring Environmental Biodiversity: Challenges, Progress, and Opportunities" gives an overview of metagenomics, with particular emphasis on the steps involved in a typical sequence-based metagenome project. It describes and discusses sample processing, sequencing technology, assembly, binning, annotation, experimental design, statistical analysis, and data storage and sharing.

Chapter "Global Nonlinear Fitness Function for Protein Structures" examines the problem of constructing fitness landscape of proteins for generating amino acid sequences that would fold into a structural fold for protein sequence design. It introduces two geometric views and proposes a formulation using mixture of nonlinear Gaussian kernel functions.

Chapter "Clinical Assessment of Disease Risk Factors Using SNP Data and Bayesian Methods" reviews new statistical methods based on Bayesian modeling, Bayesian variable partitioning, and Bayesian graphs and networks. As an example, it outlines how to use Bayesian approaches in clinical applications to perform epistasis analysis while accounting for the block-type genome structure.

Chapter "Imaging Genetics: Information Fusion and Association Techniques between Biomedical Images and Genetic Factors" covers recent studies of correlative and association analysis of medical imaging data and high-throughput genomic data. It also provides an example of parallel independent component analysis in an imaging genetic study of schizophrenia.

We have selected these topics carefully so that the book would be useful to a broad readership, including students, postdoctoral fellows, faculty and professional practitioners in bioinformatics, medical informatics, and other biomedical studies. We expect that the book can be used as a reference for upper undergraduate-level or beginning graduate-level bioinformatics/medical informatics courses.

We would like to thank the chapter authors for their excellent contributions to the book. We also would like to thank all the reviewers for their helpful comments and suggestions. This book would not have been possible without the professional support from Springer International Publishing AG, Cham.

Columbia, USA — Dong Xu
Atlanta, USA — May D. Wang
Changchun, China — Fengfeng Zhou
Shenzhen, China — Yunpeng Cai

Contents

Global Nonlinear Fitness Function for Protein Structures

Yun Xu, Changyu Hu, Yang Dai and Jie Liang

Abstract We examine the problem of constructing fitness landscape of proteins for generating amino acid sequences that would fold into an a priori determined structural fold. Such a landscape would be useful for engineering proteins with novel or enhanced biochemistry. It should be able to characterize the global fitness landscape of many proteins simultaneously, and can guide the search process to identify the correct protein sequences. We introduce two geometric views and propose a formulation using mixture of nonlinear Gaussian kernel functions. We aim to solve a simplified protein sequence design problem. Our goal is to distinguish each native sequence for a major portion of representative protein structures from a large number of alternative decoy sequences, each a fragment from proteins of different folds. The nonlinear fitness function developed discriminates perfectly a set of 440 native proteins from 14 million sequence decoys, while no linear fitness function can succeed in this task. In a blind test of unrelated proteins, the nonlinear fitness function misclassifies only 13 native proteins out of 194. This compares favorably with about 3–4 times more misclassifications when optimal linear functions are used. To significantly reduce the complexity of the nonlinear fitness function, we further constructed a simplified nonlinear fitness function using a rectangular kernel with a basis set of proteins and decoys chosen a priori. The full landscape for a large number of protein folds can be captured using only 480 native proteins and 3200 nonprotein decoys via a finite Newton method, compared to about 7000 proteins and decoys in the original nonlinear fitness function. A blind test of a simplified version of sequence design was carried out to discriminate simultaneously 428 native sequences with no significant sequence identity to any

Y. Xu · C. Hu · Y. Dai · J. Liang (✉)
Department of Bioengineering, University of Illinois, Chicago, USA
e-mail: jliang@uic.edu

Y. Xu
e-mail: yxu7@uic.edu

C. Hu
e-mail: chyhu@uic.edu

Y. Dai
e-mail: yangdai@uic.edu

D. Xu et al. (eds.), *Health Informatics Data Analysis*, Health Information Science,
DOI 10.1007/978-3-319-44981-4_1

training proteins from 11 million challenging protein-like decoys. This simplified fitness function correctly classified 408 native sequences, with only 20 misclassifications (95% correct rate), which outperforms several other statistical linear fitness functions and optimized linear functions. Our results further suggested that for the task of global sequence design, the search space of protein shape and sequence can be effectively parameterized with a relatively small number of carefully chosen basis set of proteins and decoys. For example, the task of designing 428 selected nonhomologous proteins can be achieved using a basis set of about 3680 proteins and decoys. In addition, we showed that the overall landscape is not overly sensitive to the specific choice of the proteins and decoys. The construction of fitness landscape has broad implication in understanding molecular evolution, cellular epigenetic state, and protein structures. Our results can be generalized to construct other types of fitness landscape.

Introduction

We aim to construct a global fitness function of the protein universe based on knowledge of protein structures. Such a fitness function can be used to study protein evolution and to design sequences of novel proteins. For example, the fundamental problem of protein sequence design, also called the inverse protein folding problem, aims to identify sequences compatible with a given protein fold and incompatible to alternative folds [11, 15, 61]. It has attracted considerable interests [7, 13, 25, 29, 30, 35–37, 42, 45, 64, 68, 70, 71, 92]. With successful design, one can engineer novel protein molecules with improved activities or new functions. There have been many fruitful design studies reported in the literature [1, 10, 12, 18, 41, 74].

A successful protein design strategy needs to solve two problems. First, it needs to explore both the sequence and structure spaces and efficiently generate candidate sequences. Second, a fitness or scoring function needs to identify sequences that are compatible with the desired structural fold (the "design in" principle) but are incompatible with any other competing folds (the "design out" principle) [36, 37, 92]. To achieve this, an ideal fitness function would maximize the probabilities of protein sequences taking their native fold, and reduce the probability that these sequences take any other fold. Because many protein sequences with low sequence identity can adopt the same protein fold, a full-fledged design fitness function should theoretically identify all sequences that fold into the same desired structural fold from a vast number of sequences that do fold into alternative structures, or that do not fold. Furthermore, an ideal fitness function should be able to characterize the properties of fitness landscape of many proteins simultaneously. Such a fitness function would be useful for designing novel proteins with novel functions, as well as for studying the global evolution of protein structures and protein functions.

Several scoring functions that can be used as fitness functions for protein design have been developed based on physical models. For redesigning protein cores, hydrophobicity and packing specificity are the main ingredients of the scoring functions [12]. Van der Waals interactions and electrostatics have also been incorporated for protein design [36, 37]. A combination of terms including Lennard-Jones potential, repulsion, Lazaridis–Karplus implicit solvation, approximated electrostatic interactions, and hydrogen bonds is used in an insightful computational protein design experiment [40]. Model of solvation energy based on surface area is a key component of several other design scoring functions [36, 37, 88].

A variety of empirical scoring functions based on known protein structures have also been developed for coarse-grained models of proteins. In this case, proteins are not represented in atomic details but are often represented at the residue level. Because of the coarse-grained nature of the protein representation, these scoring functions allow rapid exploration of the search space of the main factors important for proteins, and can provide good initial solutions for further refinement where models with atomistic details can be used.

Many empirical scoring functions for protein fitness were originally developed for the purposes of protein folding and structure prediction. Because the principles are somewhat similar, they are often used directly for protein design. One prominent class of empirical scoring functions is knowledge-based scoring functions, which are derived from statistical analysis of database of protein structures [49, 56, 65, 75]. Here the interactions between a pair of residues are estimated from its relative frequency in database when compared with a reference state or a null model. This approach has found many successfully applications [28, 44, 47, 49, 57, 65, 72, 73, 89]. However, there are several conceptual difficulties with this approach. These include the neglect of chain connectivity in the reference state, as well as the problematic implicit assumption of Boltzmann distribution [5, 76, 77].

An alternative approach for empirical scoring function is to find a set of parameters such that the scoring functions are optimized by some criterion, e.g., maximized score difference between native conformation and a set of alternative (or decoy) conformations [4, 14, 22, 50, 54, 76, 78, 84, 85]. This approach has been shown to be effective in protein fold recognition, where native structures can be identified from alternative conformations [54]. However, if a large number of native protein structures are to be simultaneously discriminated against a large number of decoy conformations, no such scoring functions can be found [78, 85].

There are three key steps in developing an effective empirical fitness or scoring function using optimization: (1) the functional form, (2) the generation of a large set of decoys for discrimination, and (3) the optimization techniques. The initial step of choosing an appropriate functional form is often straightforward. Empirical pairwise scoring functions are usually all in the form of weighted linear sum of interacting residue pairs. In this functional form, the weight coefficients are the parameters of the scoring function, which are optimized for discrimination. The same functional form is also used in statistical potential, where the weight coefficients are derived from database statistics. The optimization techniques that have been used include perceptron learning and linear programming [78, 85]. The objectives of optimization are often maximization of score gap between native protein and the average of decoys, or score gap between native and decoys with lowest score, or the z-score of the native protein [22, 24, 38, 39, 55].

Here we are concerned with the problem of constructing a global fitness function of proteins based on solving a simplified version of the protein design problem. We aim to develop a globally applicable scoring function for characterizing the fitness landscape of many proteins simultaneously. We would like to identify a protein sequence that is compatible with a given three-dimensional coarse-grained structure from a set of protein sequences that are taken from protein structures of different folds. We will also discuss how to proceed to develop a full-fledged fitness function that discriminates similar and dissimilar sequences adopting the same fold against all sequences that adopt different folds and sequences that do not fold. In this study, we do not address the problem of how to generate candidate template structural fold or candidate sequence by searching either the conformation space or the sequence space.

To develop an empirical fitness function that improves discrimination of native protein sequence, we examine an alternative formulation of a scoring function, in the form of mixture of nonlinear Gaussian kernel functions. We first use an optimization technique based on quadratic programming. Instead of maximizing the score gap, an objective function related to bounds of expected classification errors is optimized [8, 67, 80, 83]. Experimentation shows that the derived nonlinear function can discriminate simultaneously 440 native proteins against 14 million sequence decoys. In contrast, a perfect scoring function of weighted linear sum cannot be found using the interior point solver of linear programming following [52, 78]. We also performed blind tests for native sequence recognition. Taking 194 proteins unrelated to the 440 training set proteins, the nonlinear fitness function achieves a success rate of 93.3%. This result compares favorably with those when using optimal linear scoring function (80.9 and 73.7% success rate) and statistical potential (58.2%) [4, 57, 78].

However, this nonlinear fitness function is parameterized by about 350 native proteins and 4700 nonprotein decoys, with a rather complex form. It is computationally expensive to evaluate the fitness of a candidate sequence using this function. Although obtaining a good answer at high computational cost is acceptable for some tasks, it is difficult to incorporate a complex function in a search algorithm. It is also difficult to characterize globally the landscape properties of proteins using a complex function.

To simplify the nonlinear function for characterizing the fitness landscape of proteins, we further developed a nonlinear kernel function using a rectangular kernel, with proteins and decoys chosen a priori via a finite Newton method. The total number of native proteins and decoy conformations included in the function was reduced to about 3680. In the blind test of sequence design to discriminate 428 native sequences from 11 million challenging protein-like 18 misclassified? decoy sequences, this fitness function misclassified only 20 native sequences (correct rate 95%), which far outperforms statistical function [58] and linear optimal functions [4, 79]. It is also comparable to the results of 18 misclassifications (correct rate 91%) using the more complex nonlinear fitness function with >5000 terms [27].

Our chapter is organized as follows. We first describe: the theory and the models of linear and nonlinear functions, including the kernel and rectangle kernel models and the optimization techniques for sequence design. We then explain details of computation. Results of training and blind tests are then presented. We conclude with discussion and remarks.

Theory and Models

Modeling Protein Fitness Function

To model a protein computationally, we first need to describe its geometric shape and its sequence of amino acid residues. A protein can be represented by a d-dimensional vector $\boldsymbol{c} \in \mathbb{R}^d$. For example, a vector of number count of nonbonded contacts between different types of amino acid residues in a protein structure. In this case, the count vector $\boldsymbol{c} \in \mathbb{R}^d, d = 210$ is used as the protein descriptor. Once the structures of a protein $\boldsymbol{s}$ and its amino acid sequence $\boldsymbol{a}$ are given, the protein description f: $(\boldsymbol{s}, \boldsymbol{a}) \mapsto \mathbb{R}^d$ will give the d-dimensional vector $\boldsymbol{c}$. In the case of contact vector, f corresponds to the mapping provided by specific contact definition, e.g., two residues are in contact if their distance is below a specific cutoff threshold distance.

To develop fitness functions or scoring functions that allow the identification of sequences most compatible with a specific given coarse-grained three-dimensional structure, we can use a model analogous to the Anfinsen experiments in protein folding. We require that the native amino acid sequence $\boldsymbol{a}_N$ mounted on the native structure $\boldsymbol{s}_N$ has the best (lowest) fitness score compared to a set $\mathscr{D} = \{\boldsymbol{s}_N, \boldsymbol{a}_D\}$ of alternative sequences, called *sequence decoys*, which are taken from unrelated proteins known to fold into a different fold when mounted on the same native protein structure $\boldsymbol{s}_N$:

$$H(f(\boldsymbol{s}_N, \boldsymbol{a}_N)) < H(f(\boldsymbol{s}_N, \boldsymbol{a}_D)) \quad \text{for all } \boldsymbol{a}_D \in \mathscr{D}.$$

Equivalently, the native sequence will have the highest probability to fit into the specified native structure. This is the same principle described in [13, 45, 69]. Sometimes we can further require that the difference in fitness score must be greater than a constant $b > 0$:

$$H(f(\boldsymbol{s}_N, \boldsymbol{a}_N)) + b < H(f(\boldsymbol{s}_N, \boldsymbol{a}_D)) \quad \text{for all } (\boldsymbol{s}_D, \boldsymbol{a}_N) \in \mathscr{D}.$$

A widely used functional form for H is the weighted linear sum of pairwise contacts [49, 56, 65, 75, 78, 84]:

$$H(f(\boldsymbol{s}, \boldsymbol{a})) = H(\boldsymbol{c}) = \boldsymbol{w} \cdot \boldsymbol{c}, \tag{1}$$

where "$\cdot$" denotes inner product of vectors. As soon as the weight vector $\boldsymbol{w}$ is specified, the scoring function is fully defined. For such a linear fitness function, the basic requirement is then

$$\boldsymbol{w} \cdot (\boldsymbol{c}_N - \boldsymbol{c}_D) < 0,$$

or

$$\boldsymbol{w} \cdot (\boldsymbol{c}_N - \boldsymbol{c}_D) + b < 0, \tag{2}$$

if we require that the difference in fitness between a native protein and a decoy must be greater than a real value b. An ideal function therefore would assign the value "-1" for native structure/sequence, and the value "+1" for decoys.

Two Geometric Views of Linear Protein Potentials

There is a natural geometric view of the inequality requirement for weighted linear sum functions. A useful observation is that each of the inequalities divides the space of $\mathbb{R}^d$ into two halfs separated by a hyperplane (Fig. 1a). The hyperplane of Eq. (2) is defined by the normal vector $(\boldsymbol{c}_N - \boldsymbol{c}_D)$ and its distance $b/||\boldsymbol{c}_N - \boldsymbol{c}_D||$ from the origin. The weight vector $\boldsymbol{w}$ must be located in the half-space opposite to the direction of the normal vector $(\boldsymbol{c}_N - \boldsymbol{c}_D)$. This half-space can be written as $\boldsymbol{w} \cdot (\boldsymbol{c}_N - \boldsymbol{c}_D) + b < 0$. When there are many inequalities to be satisfied simultaneously, the intersection of the half-spaces forms a convex polyhedron [16]. If the weight vector is located in the polyhedron, all the inequalities are satisfied. Fitness functions with such a weight vector $\boldsymbol{w}$ can discriminate the native protein sequence from the set of all decoys. This is illustrated in Fig. 1a for a two-dimensional toy example, where each straight line represents an inequality $\boldsymbol{w} \cdot (\boldsymbol{c}_N - \boldsymbol{c}_D) + b < 0$ that the scoring function must satisfy.

For each native protein i, there is one convex polyhedron $\mathscr{P}_i$ formed by the set of inequalities associated with its decoys. If a scoring function can discriminate simultaneously n native proteins from a union of sets of sequence decoys, the weight vector $\boldsymbol{w}$ must be located in a smaller convex polyhedron $\mathscr{P}$ that is the intersection of the n convex polyhedra:

$$\boldsymbol{w} \in \mathscr{P} = \bigcap_{i=1}^{n} \mathscr{P}_i.$$

There is yet another geometric view of the same inequality requirements. If we now regard $(\boldsymbol{c}_N - \boldsymbol{c}_D)$ as a point in $\mathbb{R}^d$, the relationship $\boldsymbol{w} \cdot (\boldsymbol{c}_N - \boldsymbol{c}_D) + b < 0$ for all sequence decoys and native proteins requires that all points $\{\boldsymbol{c}_N - \boldsymbol{c}_D\}$ are located on one side of a different hyperplane, which is defined by its normal vector $\boldsymbol{w}$ and its distance $b/||\boldsymbol{w}||$ to the origin (Fig. 1b). Such a hyperplane exists if the origin is not contained within the convex hull of the set of points $\{\boldsymbol{c}_N - \boldsymbol{c}_D\}$ (see appendix of Ref. [27] for a proof).

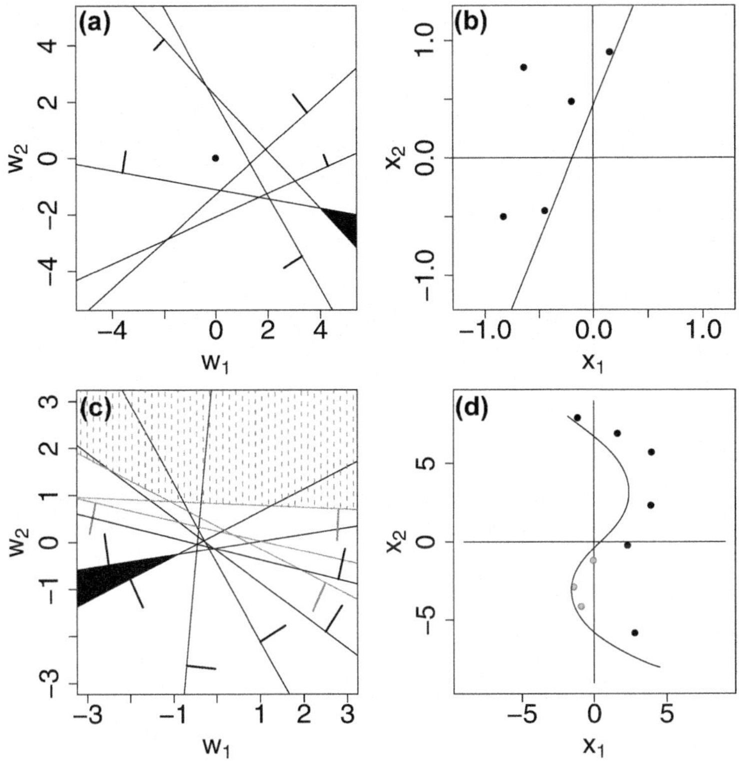

Fig. 1 Geometric views of the inequality requirement for protein scoring function. Here we use a two-dimensional toy example for illustration. **a** In the first geometric view, the space $\mathbb{R}^2$ of $w = (w_1, w_2)$ is divided into two half-spaces by an inequality requirement, represented as a *hyperplane* $w \cdot (c_N - c_D) + b < 0$. The *hyperplane*, which is a line in $\mathbb{R}^2$, is defined by the normal vector $(c_N - c_D)$, and its distance $b/||c_N - c_D||$ from the origin. In this figure, this distance is set to 1.0. The normal vector is represented by a *short line segment* whose direction points away from the *straight line*. A feasible weight vector w is located in the half-space opposite to the direction of the normal vector $(c_N - c_D)$. With the given set of inequalities represented by the *lines*, any weight vector w located in the *shaped polygon* can satisfy all inequality requirements and provides a linear scoring function that has perfect discrimination. **b** A second geometric view of the inequality requirement for linear protein scoring function. The space $\mathbb{R}^2$ of $x = (x_1, x_2)$, where $x \equiv (c_N - c_D)$, is divided into two half-spaces by the *hyperplane* $w \cdot (c_N - c_D) + b < 0$. Here the *hyperplane* is defined by the normal vector w and its distance $b/||w||$ from the origin. All points $\{c_N - c_D\}$ are located on one side of the hyperplane away from the origin, therefore satisfying the inequality requirement. That is, a linear scoring function w such as the one represented by the *straight line* in this figure can have perfect discrimination. **c** In the second toy problem, a set of inequalities are represented by a *set of straight lines* according to the first geometric view. A subset of the inequalities require that the weight vector w to be located in the *shaded convex polygon on the left*, but another subset of inequalities require that w to be located in the *dashed convex polygon on the top*. Since these two *polygons* do not intersect, there is no weight vector w that can satisfy all inequality requirements. That is, no linear scoring function can classify these decoys from native protein. **d** According to the second geometric view, no *hyperplane* can separate all points $\{c_N - c_D\}$ from the origin. But a *nonlinear curve* formed by a mixture of Gaussian kernels can have perfect separation of all vectors $\{c_N - c_D\}$ from the origin: It has perfect discrimination

The second geometric view looks very different from the first one. However, the second view is dual and mathematically equivalent to the first geometric view. In the first view, a point $\boldsymbol{c}_N - \boldsymbol{c}_D$ determined by the pair of native structure–sequence $c_N = (\boldsymbol{s}_N, \boldsymbol{a}_N)$ and decoy structure–sequence $c_D = (\boldsymbol{s}_N, \boldsymbol{a}_D)$ corresponds to a hyperplane representing an inequality. A solution weight vector $\boldsymbol{w}$ corresponds to a point located in the final convex polyhedron. In the second view, each native–decoy pair is represented as a point $\boldsymbol{c}_N - \boldsymbol{c}_D$ in $\mathbb{R}^d$, and the solution weight vector $\boldsymbol{w}$ is represented by a hyperplane separating all the points $\mathscr{C} = \{\boldsymbol{c}_N - \boldsymbol{c}_D\}$ from the origin.

Optimal Linear Fitness Function

There are many optimization methods for finding the weight vector $\boldsymbol{w}$ of linear function. The Rosenblatt perceptron method works by iteratively updating an initial weight vector $\boldsymbol{w}_0$ [54, 84]. Starting with a random vector, e.g., $\boldsymbol{w}_0 = 0$, one tests each native protein and its decoy structure. Whenever the relationship $\boldsymbol{w} \cdot (\boldsymbol{c}_N - \boldsymbol{c}_D) + b < 0$ is violated, one updates $\boldsymbol{w}$ by adding to it a scaled violating vector $\eta \cdot (\boldsymbol{c}_N - \boldsymbol{c}_D)$. The final weight vector is therefore a linear combination of protein and decoy count vectors:

$$\boldsymbol{w} = \sum \eta(\boldsymbol{c}_N - \boldsymbol{c}_D) = \sum_{N \in \mathscr{N}} \alpha_N \boldsymbol{c}_N - \sum_{D \in \mathscr{D}} \alpha_D \boldsymbol{c}_D. \tag{3}$$

Here $\mathscr{N}$ is the set of native proteins, and $\mathscr{D}$ is the set of decoys. The set of coefficients $\{\alpha_N\} \cup \{\alpha_D\}$ gives a dual form representation of the weight vector $\boldsymbol{w}$, which is an expansion of the training examples including both native and decoy structures.

According to the first geometric view, if the final convex polyhedron $\mathscr{P}$ is nonempty, there can be an infinite number of choices of $\boldsymbol{w}$, all with perfect discrimination. But how do we find a weight vector $\boldsymbol{w}$ that is optimal? This depends on the criterion for optimality. The weight vector $\boldsymbol{w}$ that minimizes the variance of score gaps between decoys and natives, $\arg_{\boldsymbol{w}} \min \frac{1}{|\mathscr{D}|} \sum (\boldsymbol{w} \cdot (\boldsymbol{c}_N - \boldsymbol{c}_D))^2 - \left[\frac{1}{|\mathscr{D}|} \sum_D (\boldsymbol{w} \cdot (\boldsymbol{c}_N - \boldsymbol{c}_D))\right]^2$, is used in Ref. [78]. Other criteria include minimizing the Z-score of a large set of native proteins, minimizing the Z-score of the native protein and an ensemble of decoys [9, 55], maximizing the ratio R between the width of the distribution of the score, and the average score difference between the native state and the unfolded ones [22, 23]. Effective linear sum scoring functions were obtained using these optimization techniques [14, 19, 22, 78, 84].

Here we describe yet another optimality criterion according to the second geometric view. We can choose the hyperplane $(\boldsymbol{w}, b)$ that separates the points $\{\boldsymbol{c}_N - \boldsymbol{c}_D\}$ with the largest distance to the origin. Intuitively, we want to characterize proteins with a region defined by the training set points $\{\boldsymbol{c}_N - \boldsymbol{c}_D\}$. It is desirable to define this region such that a new unseen point drawn from the same protein distribution as $\{\boldsymbol{c}_N - \boldsymbol{c}_D\}$ will have a high probability to fall within the defined region. Nonprotein points following a different distribution, which is assumed to be centered around the origin when no a priori information is available, will have a high probability to fall outside the defined region. In this case, we are more interested in modeling the region or support of the distribution of protein data, rather than estimating its density distribution function. For linear scoring function, regions are half-spaces defined by hyperplanes, and the optimal hyperplane $(\boldsymbol{w}, b)$ is then the one with maximal distance to the origin. This is related to the novelty detection problem and single-class support vector machine studied in statistical learning theory [67, 82, 83]. In our case, any nonprotein points will need to be detected as outliers from the protein distribution characterized by $\{\boldsymbol{c}_N - \boldsymbol{c}_D\}$. Among all linear functions derived from the same set of native proteins and decoys, an optimal weight vector $\boldsymbol{w}$ is likely to have the least amount of mislabeling. This optimal weight vector $\boldsymbol{w}$ can be found by solving the following quadratic programming problem:

$$\text{Minimize} \quad \frac{1}{2}||\boldsymbol{w}||^2 \tag{4}$$

$$\text{subject to } \boldsymbol{w} \cdot (\boldsymbol{c}_N - \boldsymbol{c}_D) + b < 0 \text{ for all } N \in \mathcal{N} \text{ and } D \in \mathcal{D}. \tag{5}$$

The solution maximizes the distance $b/||\boldsymbol{w}||$ of the plane $(\boldsymbol{w}, b)$ to the origin.

Relation to Support Vector Machines

There may exist multiple $\boldsymbol{w}$'s if $\mathcal{P}$ is not empty. We can use the formulation of a support vector machine to find a $\boldsymbol{w}$. Let all vectors $\boldsymbol{c}_N \in \mathbb{R}^d$ form a native training set and all vectors $\boldsymbol{c}_D \in \mathbb{R}^d$ form a decoy training set. Each vector in the native training set is labeled as -1 and each vector in the decoy training set is labeled as $+1$. Then solving the following support vector machine problem will provide an optimal solution to inequalities (7):

$$\begin{array}{ll} \text{Minimize} & \frac{1}{2}||\boldsymbol{w}||^2 \\ \text{subject to} & \boldsymbol{w} \cdot \boldsymbol{c}_N + b \leq -1 \\ & \boldsymbol{w} \cdot \boldsymbol{c}_D + b \geq 1. \end{array} \tag{6}$$

Note that a solution of the above problem satisfies the system of inequalities (7) below, since subtracting the second inequality from the first inequality in the constraint conditions of (6) will give us $\boldsymbol{w} \cdot (\boldsymbol{c}_N - \boldsymbol{c}_D) \leq -2 < 0$.

$$\boldsymbol{w} \cdot (\boldsymbol{c}_N - \boldsymbol{c}_D) < 0. \tag{7}$$

Equation (6) is related to the standard support vector machine formulation [8, 67, 80].

Nonlinear Scoring Function

However, it is possible that the weight vector $\boldsymbol{w}$ does not exist, i.e., the final convex polyhedron $\mathscr{P} = \bigcap_{i=1}^{n} \mathscr{P}_i$ may be an empty set. First, for a specific native protein i, there may be severe restriction from some inequality constraints, which makes $\mathscr{P}_i$ an empty set. For example, some decoys are very difficult to discriminate due to perhaps deficiency in protein representation. It would be impossible to adjust the weight vector so the native protein has a lower score than the sequence decoy. Figure 1c shows a set of inequalities represented by straight lines according to the first geometric view. A subset of inequalities (black lines) require that the weight vector $\boldsymbol{w}$ to be located in the shaded convex polygon on the left, but another subset of inequalities (green lines) require that $\boldsymbol{w}$ to be located in the dashed convex polygon on the top. Since these two polygons do not intersect, there is no weight vector that can satisfy all these inequality requirements. That is, no linear scoring function can classify all decoys from the native protein. According to the second geometric view (Fig. 1d), no hyperplane can separate all points (black and green) $\{\boldsymbol{c}_N - \boldsymbol{c}_D\}$ from the origin.

Second, even if a weight vector $\boldsymbol{w}$ can be found for each native protein, i.e., $\boldsymbol{w}$ is contained in a nonempty polyhedron, it is still possible that the intersection of n polyhedra is an empty set, i.e., no weight vector can be found that can discriminate all native proteins against the decoys simultaneously. Computationally, the question whether a solution weight vector $\boldsymbol{w}$ exists can be answered unambiguously in polynomial time [33]. When the number of decoys reaches millions, no such a weight vector can be found in a computational study [27].

A fundamental reason for this failure is that the functional form of linear sum is too simplistic. Additional descriptors of protein structures such as higher order interactions (e.g., three-body or four-body contacts) should help [6, 46, 59, 93].

Here we take an alternative approach. We still limit ourselves to pairwise contact interactions, although it can be naturally extended to include three or four body interactions [46]. We introduce a nonlinear fitness function analogous to the dual form of the linear function in Eq. (3), which takes the following form:

$$H(f(\boldsymbol{s},\boldsymbol{a})) = H(\boldsymbol{c}) = \sum_{D\in\mathscr{D}} \alpha_D K(\boldsymbol{c},\boldsymbol{c}_D) - \sum_{N\in\mathscr{N}} \alpha_N K(\boldsymbol{c},\boldsymbol{c}_N), \tag{8}$$

where $\alpha_D \geq 0$ and $\alpha_N \geq 0$ are parameters of the scoring function to be determined, and $\boldsymbol{c}_D = f(\boldsymbol{s}_N, \boldsymbol{a}_D)$ from the set of decoys $\mathscr{D} = \{(\boldsymbol{s}_N, \boldsymbol{a}_{\mathscr{D}})\}$ is the contact vector of a sequence decoy D mounted on a native protein structure $\boldsymbol{s}_N$, and $\boldsymbol{c}_N = f(\boldsymbol{s}_N, \boldsymbol{a}_N)$ from the set of native training proteins $\mathscr{N} = \{(\boldsymbol{s}_N, \boldsymbol{a}_N)\}$ is the contact vector of a native sequence $\boldsymbol{a}_N$ mounted on its native structure $\boldsymbol{s}_N$. The difference of this functional form from linear function in Eq. (3) is that a kernel function $K(\boldsymbol{x},\boldsymbol{y})$ replaces the linear term. A convenient kernel function K is

$$K(\boldsymbol{c}_i, \boldsymbol{c}_j) = e^{-\gamma||\boldsymbol{c}_i - \boldsymbol{c}_j||^2} \text{for any vectors } \boldsymbol{c}_i \text{ and } \boldsymbol{c}_j \in \mathcal{N} \cup \mathcal{D}, \tag{9}$$

where γ is a constant. The fitness function $H(\boldsymbol{c})$ can be written compactly as

$$H(\boldsymbol{c}) = \sum_{D \in \mathcal{D}} \alpha_D e^{-\gamma||\boldsymbol{c} - \boldsymbol{c}_D||^2} - \sum_{N \in \mathcal{N}} \alpha_N e^{-\gamma||\boldsymbol{c} - \boldsymbol{c}_N||^2} + b = K(\boldsymbol{c}, A) D_s \boldsymbol{\alpha} + b, \tag{10}$$

where A is the matrix of training data: $A = (\boldsymbol{c}_1^T, \ldots, \boldsymbol{c}_{|\mathcal{D}|}^T, \boldsymbol{c}_{|\mathcal{D}|+1}^T, \ldots, \boldsymbol{c}_{|\mathcal{D}|+|\mathcal{N}|}^T)^T$, and the entry $K(\boldsymbol{c}, \boldsymbol{c}_j)$ of $K(\boldsymbol{c}, A)$ is $e^{-\gamma||\boldsymbol{c} - \boldsymbol{c}_j||^2}$. D_s is the diagonal matrix with +1 and -1 along its diagonal representing the membership class of each point $A_i = \boldsymbol{c}_i^T$. Here $\boldsymbol{\alpha}$ is the coefficient vector: $\boldsymbol{\alpha} = (\alpha_1, \ldots, \alpha_{|\mathcal{D}|}, \alpha_{|\mathcal{D}|+1}, \ldots, \alpha_{|\mathcal{D}|+\mathcal{N}|})^T$.

Intuitively, the fitness landscape has smooth Gaussian hills of height α_D centered on location $\boldsymbol{c}_D$ of decoy contact vector $D \in \mathcal{D}$, and has smooth Gaussian cones of depth α_N centered on the location $\boldsymbol{c}_N$ of native contact vector $N \in \mathcal{N}$. Ideally, the value of the fitness function will be -1 for contact vectors $\boldsymbol{c}_N$ of native proteins, and will be +1 for contact vectors $\boldsymbol{c}_D$ of decoys.

Optimal Nonlinear Fitness Function

To obtain such a nonlinear function, our goal is to find a set of parameters $\{\alpha_D, \alpha_N\}$ such that $H(\boldsymbol{c})$ has fitness value close to -1 for native proteins, and has fitness values close to +1 for decoys. There are many different choices of $\{\alpha_D, \alpha_N\}$. We use an optimality criterion developed in statistical learning theory [8, 66, 81]. First, we note that we have implicitly mapped each protein and decoy from $\mathbb{R}^d, d = 210$ to another high-dimensional space where the scalar product of a pair of mapped points can be efficiently calculated by the kernel function $K(.,.)$. Second, we find the hyperplane of the largest margin distance separating proteins and decoys in the space transformed by the nonlinear kernel [8, 66, 81, 83]. That is, we search for a hyperplane with equal and maximal distance to the closest native protein sequence and the closest decoys. Such a hyperplane has good performance in discrimination [81]. It can be found by using support vector machine to obtain the parameters $\{\alpha_D\}$ and $\{\alpha_N\}$ from solving the following primal form of quadratic programming problem:

$$\begin{array}{ll} \min\limits_{\boldsymbol{\alpha} \in R_+^m, b \in R, \xi \in R^m} & \frac{C}{2} \boldsymbol{e} \cdot \xi + \frac{1}{2} \boldsymbol{\alpha} \cdot \boldsymbol{\alpha} \\ \text{subject to} & D_s(K(A, A) D_s \boldsymbol{\alpha} + b\boldsymbol{e}) + \xi \geq \boldsymbol{e} \\ & \xi \geq 0, \end{array} \tag{11}$$

where m is the total number of training points: $m = |\mathcal{D}| + |\mathcal{N}|$, C is a regularizing constant that limits the influence of each misclassified conformation [8, 66, 81, 83], and the $m \times m$ diagonal matrix of signs D_s with +1 or -1 along its diagonal indicating the membership of each point A_i in the classes +1 or −1; and $\mathbf{e}$ is an m-vector with 1 at each entry. The variable ξ_i is a measurement of error for each input vector with respect to the solution: $\xi_i = 1 + y_i H(\boldsymbol{c}_i)$, where $y_i = -1$ if i is a native protein, and $y_i = +1$ if i is a decoy.

Rectangle Kernel and Reduced Support Vector Machine (RSVM)

The use of nonlinear kernels on large datasets typically demands a prohibiting size of the computer memory in solving the potentially enormous unconstrained optimization problem. Moreover, the representation of the landscape surface using a large data set requires costly storage and computing time for the evaluation of a new unseen contact vector $\mathbf{c}$. To overcome these difficulties, the reduced support vector machines (RSVM) developed by Lee and Mangasarian [43] use a very small random subset of the training set to build a rectangular kernel matrix, instead of the use of the conventional $m \times m$ kernel matrix $K(A, A)$ in Eq. (11) for a training set of m examples. This model can achieve about 10% improvement on test accuracy over conventional support vector machine with random data sets of sizes between 1 and 5% of the original data [43]. The small subset can be regarded as a basis set in our study. Suppose that the number of contact vectors in our basis set is $\bar{m}$, with $\bar{m} \ll m$. We denote $\bar{A}$ as an $\bar{m} \times d$ matrix, and each contact vector from the basis set is represented by a row vector of $\bar{A}$. The resulting kernel matrix $K(A, \bar{A})$ from A and $\bar{A}$ has size $m \times \bar{m}$. Each entry of this rectangular kernel matrix is calculated by $K(\boldsymbol{c}_i, \bar{\mathbf{c}}_j)$, where $\boldsymbol{c}_i^T$ and $\bar{\mathbf{c}}_j^T$ are rows from A and $\bar{A}$, respectively. The RSVM is formulated as the following quadratic program:

$$\begin{array}{ll} \min\limits_{\bar{\boldsymbol{\alpha}} \in \mathbb{R}_+^{\bar{m}}, b \in \mathbb{R}, \xi \in \mathbb{R}^{\bar{m}}} & \frac{C}{2} \boldsymbol{\xi} \cdot \boldsymbol{\xi} + \frac{1}{2} (\bar{\boldsymbol{\alpha}} \cdot \bar{\boldsymbol{\alpha}} + b^2) \\ \text{subject to} & D_s (K(A, \bar{A}) \bar{D}_s \bar{\boldsymbol{\alpha}} + b\boldsymbol{e}) + \boldsymbol{\xi} \geq \boldsymbol{e} \\ & \boldsymbol{\xi} \geq 0, \end{array} \tag{12}$$

where $\bar{D}_s$ is the $\bar{m} \times \bar{m}$ diagonal matrix with +1 or −1 along its diagonal, indicating the membership of each point $\bar{A}_i$ in the classes +1 or −1; and $\boldsymbol{e}$ is an m-vector with 1 at each entry. As shown in [43], the zero level set surface of the fitness function is given by

$$H(\boldsymbol{c}) = K(\boldsymbol{c}, \bar{A}) \bar{D}_s \bar{\boldsymbol{\alpha}} + b = \sum_{\boldsymbol{c}_D \in \bar{A}} \bar{\alpha}_D e^{-\gamma \|\boldsymbol{c} - \boldsymbol{c}_D\|^2} - \sum_{\boldsymbol{c}_N \in \bar{A}} \bar{\alpha}_N e^{-\gamma \|\boldsymbol{c} - \boldsymbol{c}_N\|^2} + b = 0, \tag{13}$$

where $(\bar{\boldsymbol{\alpha}}, b) \in \mathbb{R}^{\bar{m}+1}$ is the unique solution to (12). This surface discriminates native proteins against decoys. Besides the rectangular kernel matrix, the use of 2-norm for the error $\boldsymbol{\xi}$ and an extra term b^2 in the objective function of (12) distinguishes this formulation from conventional support vector machine.

Smooth Newton Method

In order to solve Eq. (12) efficiently, an equivalent unconstrained nonlinear program based on the implicit Lagrangian formulation of (12) was proposed in [20], which can be solved using a fast Newton method. We modified the implicit Lagrangian formulation and obtain the unconstrained nonlinear program for the imbalanced RSVM in Eq. (12). The Lagrangian dual of (12) is now [51]:

$$\min_{\bar{\boldsymbol{\alpha}} \in \mathbb{R}^{\bar{m}}_{+}} \frac{1}{2} \bar{\boldsymbol{\alpha}} \cdot (Q + \bar{D}_s (K(A,\bar{A})^T K(A,\bar{A}) + \boldsymbol{e}\boldsymbol{e}^T \bar{D}_s) \bar{\boldsymbol{\alpha}} - \boldsymbol{e} \cdot \bar{\boldsymbol{\alpha}}, \tag{14}$$

where $Q = I/C \in \mathbb{R}^{\bar{m} \times \bar{m}}$, and $I \in \mathbb{R}^{\bar{m} \times \bar{m}}$ is a unit matrix. Note that $\mathbb{R}^{\bar{m}}_{+}$ is the set of nonnegative $\bar{m}$-vectors. Following [20], an equivalent unconstrained piecewise quadratic minimization problem of the above positively constrained optimization can be derived as follows:

$$\begin{aligned} & \min_{\bar{\boldsymbol{\alpha}} \in \mathbb{R}^{\bar{m}}} L(\bar{\boldsymbol{\alpha}}) \\ = \quad & \min_{\bar{\boldsymbol{\alpha}} \in \mathbb{R}^{\bar{m}}} \tfrac{1}{2} \bar{\boldsymbol{\alpha}} \cdot Q\bar{\boldsymbol{\alpha}} - \boldsymbol{e} \cdot \bar{\boldsymbol{\alpha}} + \tfrac{1}{2}\beta(\|(-\beta\bar{\boldsymbol{\alpha}} + Q\bar{\boldsymbol{\alpha}} - \boldsymbol{e})_{+}\|^2 - \|Q\bar{\boldsymbol{\alpha}} - \boldsymbol{e}\|^2). \end{aligned} \tag{15}$$

Here, β is a sufficiently large but bounded positive parameter to ensure that the matrix $\beta I - Q$ is positive definite, where the plus function $(\cdot)_{+}$ replaces negative components of a vector by zeros. This unconstrained piecewise quadratic problem can be solved by the Newton method in a finite number of steps [20]. The Newton method requires the information of the gradient vector $\nabla L(\bar{\boldsymbol{\alpha}}) \in \mathbb{R}^{\bar{m}}$ and the generalized Hessian $\partial^2 L(\bar{\boldsymbol{\alpha}}) \in \mathbb{R}^{\bar{m} \times \bar{m}}$ of $L(\bar{\boldsymbol{\alpha}})$ at each iteration. They can be calculated using the following formula [20]:

$$\begin{aligned} \nabla L(\bar{\boldsymbol{\alpha}}) &= (Q\bar{\boldsymbol{\alpha}} - \boldsymbol{e}) + \tfrac{1}{\beta}(Q - \beta I)((Q - \beta I) - \boldsymbol{e})_{+} - \tfrac{1}{\beta} Q(Q\bar{\boldsymbol{\alpha}} - \boldsymbol{e}) \\ &= \tfrac{(\beta I - Q)}{\beta}((Q\bar{\boldsymbol{\alpha}} - \boldsymbol{e}) - ((Q - \beta I)\bar{\boldsymbol{\alpha}} - \boldsymbol{e})_{+}), \end{aligned} \tag{16}$$

and

$$\partial^2 L(\bar{\boldsymbol{\alpha}}) = \frac{\beta I - Q}{\beta}(Q + diag((Q - \beta I)\bar{\boldsymbol{\alpha}} - \boldsymbol{e})_{*}(\beta I - Q)), \tag{17}$$

where $diag(\cdot)$ denotes a diagonal matrix and $(\boldsymbol{\alpha})_{*}$ denotes the step function, i.e., $(\alpha_i)_{*} = 1$ if $\alpha_i > 0$; and $(\alpha_i)_{*} = 0$ if $\alpha_i \leq 0$.

The main step of the Newton method is to solve iteratively the system of linear equations

$$-\nabla L(\bar{\boldsymbol{\alpha}}^i) + \partial^2 L(\bar{\boldsymbol{\alpha}}^i)(\bar{\boldsymbol{\alpha}}^{i+1} - \bar{\boldsymbol{\alpha}}^i) = \mathbf{0}, \tag{18}$$

for the unknown vector $\bar{\boldsymbol{\alpha}}^{i+1}$ with given $\bar{\boldsymbol{\alpha}}^i$.

We present below the algorithm, whose convergence was proved in [20]. We denote $\partial^2 L(\bar{\boldsymbol{\alpha}}^i)^{-1}$ as the inverse of the Hessian $\partial^2 L(\bar{\boldsymbol{\alpha}}^i)$.

Start with any $\bar{\boldsymbol{\alpha}}^0 \in \mathbb{R}^{\bar{m}}$. For $i = 0, 1\ldots$:

(i) Stop if $\nabla L(\bar{\boldsymbol{\alpha}}^i - \partial^2 L(\bar{\boldsymbol{\alpha}}^i)^{-1} \nabla L(\bar{\boldsymbol{\alpha}}^i)) = 0$.

(ii) $\bar{\boldsymbol{\alpha}}^{i+1} = \bar{\boldsymbol{\alpha}}^i - \lambda_i \partial^2 L(\bar{\boldsymbol{\alpha}}^i)^{-1} \nabla L(\bar{\boldsymbol{\alpha}}^i) = \bar{\boldsymbol{\alpha}}^i + \lambda_i \boldsymbol{d}_i$, where $\lambda_i = \max\{1, \frac{1}{2}, \frac{1}{4}, \cdots\}$ is the Armijo step size [60] such that

$$L(\bar{\boldsymbol{\alpha}}^i) - L(\bar{\boldsymbol{\alpha}}^i + \lambda_i \boldsymbol{d}^i) \geq -\delta\lambda_i \nabla L(\bar{\boldsymbol{\alpha}}^i) \cdot \boldsymbol{d}^i, \tag{19}$$

for some $\delta \in (0, \frac{1}{2})$, and $\boldsymbol{d}^i$ is the Newton direction

$$\boldsymbol{d}^i = \bar{\boldsymbol{\alpha}}^{i+1} - \bar{\boldsymbol{\alpha}}^i = -\partial^2 L(\bar{\boldsymbol{\alpha}}^i)^{-1} \nabla L(\bar{\boldsymbol{\alpha}}^i), \tag{20}$$

obtained by solving (18).

(iii) $i = i + 1$. Go to (i).

Computational Procedures

Protein Data

Protein Data for Linear and Full Nonlinear Fitness Function.

Following reference [86], we use protein structures contained in the WHATIF database [21] in this study. WHATIF database contains a representative set of sequence-unique protein structures generated from X-ray crystallography. Structures selected for this study all have pairwise sequence identity <30%, R-factor <0.21, and resolution <2.1 Å. This provides a good representative set of all known protein structures.

Specifically, we use a list of 456 proteins compiled from the 1998 release (WHATIF98) of the WHATIF database [85]. There are 192 proteins with multiple chains in this dataset. Some of them have extensive interchain contacts. For these proteins, it is possible that their conformations may be different if there are no interchain contacts present. Thirteen protein chains are removed because they all have extensive interchain contacts. We further remove three proteins because each has >10% of residues missing with no coordinates in the Protein Data Bank file. The remaining set of 440 proteins are then used as training set for developing functions. Using the threading method described in section "Contact Maps and Sequence Decoys", we generated a set of 14,080,766 sequence decoys.

Protein Data for Simplified Nonlinear Fitness Function

For constructing the simplified nonlinear fitness function using the rectangular kernel, we used a list of 1515 protein chains compiled from the PISCES server [87]. Protein chains in this data set have pairwise sequence identity <20%, a resolution ≤ 1.6 Å, and an R-factor ≤ 0.25. We removed incomplete proteins (i.e., those with missing residues), and proteins with uncertain residues, as well as proteins with fewer than 46 and more than 500 amino acids. In addition, we removed protein chains with more than 30% extensive interchain contacts. The remaining set of 1228 proteins are then randomly divided into two sets. One set includes 800 proteins and the other one includes 428 proteins. Using the sequence threading method, we generated 36,823,837 nonprotein decoys, together with 800 native proteins as the training set, and 11,144,381 decoy nonproteins with 428 native proteins as the test set.

As there is a wide range of protein chain length in this dataset, we normalize the number of contacts for each type of pairwise contact of a protein using Eq. (21) in the study of nonlinear fitness function using rectangle kernel. This equation is obtained from a linear regression on the relationship between the number of total contacts and the length of the protein,

$$N_{\text{contacts}} = 3.090 \cdot L_{\text{protein}} - 76.182, \tag{21}$$

where N_{contacts} is the number of contacts for a protein, and L_{protein} is the number of the protein residues.

Contact Maps and Sequence Decoys

Alpha Contact Maps

Because protein molecules are formed by thousands of atoms, their shapes are complex. We use the count vector of pairwise contact interactions. Here contacts are derived from the edge simplices of the alpha shape of a protein structure [17, 47, 48]. These edge simplices represent nearest-neighbor interactions that are in physical contacts. They encode precisely the same contact information as a subset of the edges in the Voronoi diagram of the protein molecule. These Voronoi edges are shared by two interacting atoms from different residues, but intersect with the body of the molecule modeled as the union of atom balls. We refer to references [17, 47, 48] for further theoretical and computational details.

Generating Sequence Decoys by Threading

We use the gapless threading method to generate a large number of decoys [32, 50, 59]. We thread the sequence of a larger protein through the structure of a smaller protein, and obtain sequence decoys by mounting a fragment of the sequence of the large protein to the full structure of the small protein. We therefore have for each

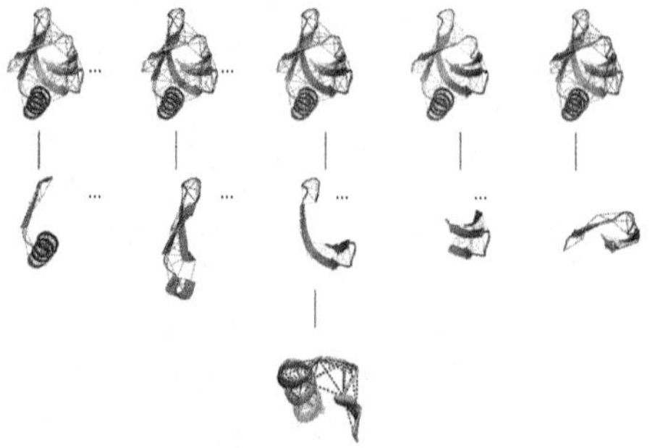

Fig. 2 Decoy generation by gapless threading. Sequence decoys can be generated by threading the sequence of a larger protein to the structure of an unrelated smaller protein. As an illustration, here the sequences of different portions of the larger protein structure (*top*), which have different native substructures (*middle*), are threaded onto the same unrelated smaller protein (*bottom*). Specifically, for a small protein of length n and a large protein of length N, we first take the subsequence from 1 to n of the larger protein and map it onto the structure of the small protein. We then start at position 2 and take the subsequence from 2 to $n+1$. This is repeated until the last decoy is generated by taking the subsequence from $N-n+1$ to N from the larger protein. Altogether, we can obtain a total of $N-n+1$ decoys from this protein pair

native protein $(\boldsymbol{s}_N, \boldsymbol{a}_N)$ a set of sequence decoys $(\boldsymbol{s}_N, \boldsymbol{a}_D)$ (Fig. 2). Because all native contacts are retained in this case, sequence decoys obtained by gapless threading are challenging.

Learning Linear Fitness Function

For comparison, we have also developed an optimal linear fitness function following the method and computational procedure described in reference [78]. We apply the interior point method as implemented in BPMD package by Mészáros [53] to search for a weight vector $\boldsymbol{w}$. We use two different optimization criteria as described in Ref. [78]. The first is

$$\begin{array}{cc} \text{Identify} & \boldsymbol{w} \\ \text{subject to } \boldsymbol{w}\cdot(\boldsymbol{c}_N-\boldsymbol{c}_D)<\varepsilon & \text{and } |w_i|\le 10, \end{array}$$

where w_i denotes the i-th component of weight vector $\boldsymbol{w}$, and $\varepsilon = 1\times 10^{-6}$. Let $\mathscr{C} = \{c_N - c_D\}$, and $|\mathscr{C}|$ the number of decoys. The second optimization criterion is

$$\begin{array}{ll} \text{Minimize} & \min \frac{1}{|\mathscr{C}|}\sum(\boldsymbol{w}\cdot(c_N-c_D))^2 - \left[\frac{1}{|\mathscr{C}|}\sum(\boldsymbol{w}\cdot(\boldsymbol{c}_N-\boldsymbol{c}_D))\right]^2 \\ \text{subject to} & \boldsymbol{w}\cdot(\boldsymbol{c}_N-\boldsymbol{c}_D)<\varepsilon. \end{array}$$

Learning Full Nonlinear Fitness Function

We use SVMlight (http://svmlight.joachims.org/) [31] with Gaussian kernels and a training set of 440 native proteins plus 14,080,766 decoys to obtain the optimized parameter $\{\alpha_N, \alpha_D\}$. The regularization constant C takes the default value, which is estimated from the training set $\mathcal{N} \cup \mathcal{D}$:

$$C = |\mathcal{N} \cup \mathcal{D}|^2 / \left[\sum_{\boldsymbol{x} \in \mathcal{N} \cup \mathcal{D}} \sqrt{K(\boldsymbol{x},\boldsymbol{x}) - 2 \cdot K(\boldsymbol{x},0) + K(0,0)} \right]^2. \quad (22)$$

Since we cannot load all 14 millions decoys into computer memory simultaneously, we use a heuristic strategy for training. Similar to the procedure reported in [78], we first randomly selected a subset of decoys that fits into the computer memory. Specifically, we pick every 51st decoy from the list of 14 million decoys. This leads to an initial training set of 276,095 decoys and 440 native proteins. An initial protein fitness function is then obtained. Next the scores for all 14 million decoys and all 440 native proteins are evaluated. Three decoy sets were collected based on the evaluation results: the first set contains the violating decoys which have lower score than the native structures; the second set contains decoys with the lowest absolute score, and the third set contains decoys that participate in $H(\boldsymbol{c})$ as identified in previous training process. The union of these three subsets of decoys is then combined with the 440 native proteins as the training set for the next iteration of learning. This process is repeated until the score difference to native protein for all decoys is greater than 0.0. Using this strategy, the number of iterations typically is between 2 and 10. During the training process, we set the cost factor j in SVMlight to 120, which is the factor training errors on native proteins, outweighs training errors on decoys.

The value of σ^2 for the Gaussian kernel $K(\boldsymbol{x},\boldsymbol{y}) = e^{-||\boldsymbol{x}-\boldsymbol{y}||^2/2\sigma^2}$ is chosen by experimentation. If the value of σ^2 is too large, no parameter set $\{\alpha_N, \alpha_D\}$ can be found such that the fitness scoring function can perfectly classify the 440 training proteins and their decoys, i.e., the problem is unlearnable. If the value of σ^2 is too small, the performance in blind test will deteriorate. The final design fitness function is obtained with σ^2 set to 416.7.

Learning Simplified Nonlinear Fitness Function

Selection of matrix A for iterative training

We used only a subset of the 36 million decoys and native structures so they could fit into the computer memory during training. These structures formed the data matrix A, which was used to construct the kernel matrix $K(A,\bar{A})$. We used a heuristic iterative approach to construct matrices A and $\bar{A}$ during each iteration.

Initially, we randomly selected 10 decoys for each of the j-th native protein from the set of decoys $\mathscr{D}_j$. We have then $m \approx 8000$ decoys for the 800 native proteins. We further chose only 1 decoy from the selected 10 decoys for each native protein j. These 800 decoys were combined with the 800 native proteins to form the initial matrix A. The contact vectors of a subset of 480 native proteins (60% of the original 800 proteins) and 320 decoys (40% of the 800 selected decoys) were then randomly chosen to form $\bar{A}$. An initial fitness function $H(\boldsymbol{c})$ was then obtained using A and $\bar{A}$. The fitness values of all 36 million decoys and the 800 native proteins were then evaluated using $H(\boldsymbol{c})$. We further used two iterative strategies to improve upon the fitness function $H(\boldsymbol{c})$.

[Strategy 1] In the i-th iteration, we selected the subset of misclassified decoys from $\mathscr{D}_j$ associated with the j-th native protein and sorted them by their fitness value in descending order, so the misclassified decoys with least violation, namely, negative but smallest absolute values in $H(\boldsymbol{c})$, are on the top of the list. If there are fewer than 10 misclassified decoys, we add top decoys that were misclassified in the previous iteration for this native protein, if they exist, such that each native protein has 10 decoys.

A new version of the matrix A was then constructed using these 8000 decoys and the corresponding 800 native proteins. To obtain the updated $\bar{A}$, from these 8800 contact vectors, we randomly selected 480 native proteins (60%) and 3200 unpaired decoy nonproteins (40%) to form $\bar{A}$.

The iterative training process was then repeated until there was no improvement in the classification of the 36 million decoys and the 800 native proteins from the training set. Typically, the number of iterations was about 10. In subsequent studies, we experimented with different percentages of selected decoys, ranging from 10 to 100% to examine the effect of the size of $\bar{A}$ on the effectiveness of the fitness function $H(\boldsymbol{c})$.

[Strategy 2] In the i-th iteration, we selected the top 10 correctly classified decoys sorted by their fitness value in ascending order for each native protein, namely, those correctly classified decoy with positive but smallest absolute values are selected. These contact vectors of 8000 selected decoys are combined with the 800 native proteins to form the new data matrix A.

To construct $\bar{A}$, we first selected the most challenging native proteins by taking the top 80 correctly classified native proteins (10%) sorted by their fitness value in descending order, namely, those that are negative but with smallest absolute values in $H(\boldsymbol{c})$. We then randomly took 400 native proteins (50%) from the rest of the native protein set, so altogether we have 480 native proteins (60%). Similarly, we selected the top one decoy that is most challenging from the 10 chosen decoys in A for each native protein, namely, the top decoy that is correctly classified with positive but smallest value of $H(\boldsymbol{c})$. We then randomly selected three decoys for each native protein from the remaining decoys in A to obtain 3200 decoy non-proteins (40%). The matrix $\bar{A}$ is then constructed from the selected 480 native proteins and 3200 decoy nonproteins. The iterative training process was repeated until there was no improvement in classification of the 36 million decoys and 800 native proteins in the training set. Typically, the number of iteration was about 5.

In the subsequent studies, we evaluated our method with different choices of challenging native proteins. The selection ranges from the top 10 to 60% most challenging native proteins. The choice of the challenging decoys was also varied, where we experimented with choosing the top one to the top four most challenging decoys for each native protein, while the number randomly selected decoys varies from three to zero.

Learning parameters

There are two important parameters: the constant γ in the kernel function $e^{-\gamma||c_i - c||^2}$, and the cost factors C, which is used during training so errors on positive examples were adjusted to outweigh errors on negative examples. Our experimentation showed that $\gamma = 5.0 \times 10^{-5}$ and $C = 1.0 \times 10^4$ were reasonable choices.

Results

Linear Fitness Functions

To search for the optimal weight vector $\boldsymbol{w}$ for the linear fitness function, we used linear programming solver based on interior point method as implemented in BPMD by Mészáros [53]. After generating 14,080,766 sequence design decoys for the 440 proteins in the training set, we searched for an optimal $\boldsymbol{w}$ that can discriminate native sequences from decoy sequences, namely, parameters $\boldsymbol{w}$ for $H(\boldsymbol{s}, \boldsymbol{a}) = \boldsymbol{w} \cdot \boldsymbol{c}$, such that $\boldsymbol{w} \cdot \boldsymbol{c}_N < \boldsymbol{w} \cdot \boldsymbol{c}_D$ for all sequences. However, we failed to find a feasible solution for the weight vector $\boldsymbol{w}$. That is, no $\boldsymbol{w}$ exists capable of discriminating perfectly 440 native sequences from the 14 million decoy sequences. We repeated the same experiment using a larger set of 572 native proteins from reference [78] and 28,261,307 sequence decoys. The result was also negative.

Full Nonlinear Fitness Function

We used the set of 440 native proteins and 14 million decoys to derive nonlinear kernel fitness functions. We succeeded in finding a function in the form of Eq. (8) that can discriminate all 440 native proteins from 14 million decoys.

Unlike statistical scoring functions where each native protein in the database contributes to the empirical scoring function, only a subset of native proteins contribute and have $\alpha_N \neq 0$. In addition, a small fraction of decoys also contribute to the fitness function. Table 1 lists the details of the fitness function, including the numbers of native proteins and decoys that participate in Eq. (8). These numbers

Table 1 Derivation of kernel fitness function

		Design Scoring Function
		$\sigma^2 = 416.7$
Num. of Vectors	Natives	220
	Decoys	1685
Range of Score Values	Natives	$0.9992 \sim 4.598$
	Decoys	$-9.714 \sim 0.7423$
Range of Smallest Score Gap		$0.2575 \sim 11.53$

Details of derivation of nonlinear kernel design scoring functions. The numbers of native proteins and decoys with nonzero α_i entering the scoring function are listed. The range of the score values of natives and decoys are also listed, as well as the range of the smallest gaps between the scores of the native protein and decoy. Details for nonlinear kernel folding scoring function are also listed

represent about 50% of native proteins and $<0.1\%$ of decoys from the original training data.

Discrimination Tests for Sequence Design Using Full Nonlinear Fitness Function.

Blind test in discriminating native proteins from decoys for an independent test set is essential to assess the effectiveness of design fitness or scoring functions. To construct such a test set, we first take the entries in WHATIF99 database that are not present in WHATIF98. After eliminating proteins with chain length less than 46 residues, we obtain a set of 201 proteins. These proteins all have <30% sequence identities with any other sequence in either the training set or the test set proteins. Since 139 of the 201 test proteins have multiple chains, we use the same criteria applied in training set selection to exclude 7 proteins with extensive interchain contacts, or with >10% residues missing in the PDB files. This leaves a smaller set of 194 test proteins. Using gapless threading, we generate a set of 3,096,019 sequence decoys from the set of 201 proteins. This is a superset of the decoy set generated using 194 proteins.

To test design fitness functions for discriminating native proteins from sequence decoys in both the 194 and the 201 test sets, we take the sequence $\boldsymbol{a}$ from the conformation–sequence pair $(\boldsymbol{s}_N, \boldsymbol{a})$ for a protein with the lowest score as the predicted sequence. If it is not the native sequence $\boldsymbol{a}_N$, the discrimination failed and the design fitness function does not work for this protein.

For comparison, we also test the discrimination results of optimal linear scoring function taken as reported in Ref. [78], as well as the statistical potential developed by Miyazawa and Jernigan. Here we use the contact definition reported in [78], that is, two residues are declared to be in contact if the geometric centers of their side chains are within a distance of 2.0–6.4 Å.

The nonlinear design fitness function capable of discriminating all of the 440 native sequences works well for the test set (Table 2). It succeeded in correctly identifying 93.3% (181 out of 194) of native sequences in the independent test set of 194 proteins. This compares favorably with results obtained using optimal linear

Table 2 Number of misclassification compared with other methods

Method	Training set 800/36 M	Training set 440/14 M	Test set 428/11 M	Test set 201/3 M
Nonlinear function	4/988	NA	20/218	NA
Tobi et al.	NA	192/39,583	NA	44/53,137
Bastolla et al.	NA	134/47,750	NA	58/29,309
Miyazawa and Jernigan	NA	173/229,549	NA	87/80,716

The number of misclassifications using simplified nonlinear fitness function, optimal linear scoring function taken as reported in [4, 79], and Miyazawa–Jernigan statistical potential [58] for both native proteins and decoys (separated by "/") in the test set and the training set. The simplified nonlinear function is formed using a basis set of 3680 (480 native + 3200 decoy) contact vectors derived using Strategy 2

folding scoring function taken as reported in [78], which succeeded in identifying 80.9% (157 out of 194) of this test set. It also has better performance than optimal linear scoring function based on calculations using parameters reported in reference [4], which succeeded in identifying 73.7% (143 out of 194) of proteins in the test set. The Miyazawa–Jernigan statistical potential succeeded in identifying 113 native proteins out of 194 (success rate 58.2%).

Running time

The evaluation of the nonlinear fitness function requires more computation than linear function, but the time requirement is modest: on an AMD Athlon MP1800+ machine of 1.54 GHz clock speed with 2 GB memory, we can evaluate the fitness function for 8130 decoys per minute.

Results of Simplified Nonlinear Fitness Function

Performance in discrimination

We used the set of 428 native proteins and 11,144,381 decoys for testing the designed fitness function. We took the sequence $\boldsymbol{a}$ for a protein such that $\boldsymbol{c} = f(s_N, \boldsymbol{a})$ has the best fitness value as the predicted sequence. If it is not the native sequence $\boldsymbol{a}_N$, then the design failed and the fitness function did not work for this protein.

The simplified nonlinear fitness function for protein design we obtained is capable of discriminating 796 of the 800 native sequences (Table 2). It also succeeded in correctly identifying 95% (408 out of 428) of the native sequences in the independent test set. Results for other methods were taken from literature obtained using much smaller and less challenging data set. Overall, the performance of our method is better than results obtained using the optimal linear scoring function taken as reported in [79] and in [4], which succeeded in identifying 78% (157 out of 201) and 71% (143 out of 201) of the test set, respectively. Our results are also better than the Miyazawa–Jernigan statistical potential [58] (success rate 58%, 113 out of 201). This performance is also comparable with the full nonlinear fitness function, with >5000 terms [27], which succeeded with a correct rate of 91% (183 out of 201).

Effect of the size of the basis set $\bar{A}$ using Strategy 1

The matrix $\bar{A}$ contains both proteins and decoys from A and its size is important in discrimination of native proteins from decoys. We examined the effects of different sizes of $\bar{A}$ using Strategy 1. For a data matrix A consisting of 800 native proteins and 8000 sequence decoys derived following the procedure described earlier, we tested different choices of $\bar{A}$ on the performance of discrimination. With the data matrix A, we fixed the selection of the 480 native proteins (60%) and experimented with random selection of different numbers of decoys, ranging from 800 (10%) to 8000 (100%) to form different $\bar{A}$s.

The results of classifying both the training set of 800 native proteins with 36 million decoys and the test set of 428 native proteins with 11 million decoys are shown in Table 3. When 60% (480) native proteins and 100% (8000) decoys are included, there are only 5 native proteins misclassified in the training set and 24 native proteins in the test set.

Effect of the size of the preselection of dataset using Strategy 2

We also examined the effects of different choices in constructing matrix $\bar{A}$ using Strategy 2. We varied our selection of the most challenging native proteins from the top 10 to 60%, and varied selection of the most challenging decoys from the top one to the top four decoys for each native protein. Results are shown in Table 4. We found that the performances of the discrimination of both the training set and test set have little changes when either native proteins selection rate is changed from 10 to 60%, or decoys selection rate is changed from the top 1 to the top 4. Overall, these results suggest that a fitness function with good discrimination can be achieved with about 480 native proteins and 3200 decoys, along with 400 preselected native proteins and 800 preselected top-1 decoys. Our final fitness function used in Table 2 is constructed using a basis set of 3680 contact vectors. The average number of iterations is about 5 using Strategy 2, which is much faster than Strategy 1.

Overall, using Strategy 2 leads to overall better performance compared to using Strategy 1 (Table 4 vs. Table 3). That is, the fitness function formed by preselecting the top 1 decoys and top 50% native proteins using Strategy 2 works well to discriminating native proteins from decoys. Furthermore, our method is robust. The overall performance using either Strategy 1 or Strategy 2 is stable when decoy selection rate changes from 5 to 90%.

Discrimination against a different decoy set

We further examine how well decoys generated by a different approach can be discriminated using the nonlinear fitness function. We selected 799 training proteins and 428 test proteins for this further test. Figure 4a shows the length distribution of

Table 3 Effects of the size of basis set $\bar{A}$ on performance of discrimination using Strategy 1

Select decoys rate (%)	Iteration	Training set Native/Decoy 800/36 M	F_β	Test set Native/Decoy 428/11 M	F_β
0	4	21/1374	0.958	26/387	0.931
2	5	19/1029	0.964	27/219	0.933
5	5	17/1303	0.963	21/317	0.944
8	5	13/1246	0.969	23/274	0.941
10	5	14/922	0.972	24/216	0.940
20	6	16/902	0.969	28/250	0.930
30	6	10/1037	0.975	29/304	0.926
40	10	16/812	0.970	27/199	0.933
50	10	13/1112	0.971	25/269	0.936
60	12	15/802	0.972	27/237	0.932
70	9	13/947	0.973	24/256	0.939
80	8	11/1078	0.973	28/278	0.929
90	9	12/690	0.977	27/170	0.934
100	5	5/2681	0.962	24/609	0.931

The number of misclassifications of both native proteins and decoys (separated by "/") with select native proteins rate 60% in both training set and test set is listed. Misclassifications as well as the F_β scores in two tests using different numbers of native proteins and decoys are listed (see text for details). Here the F_β score is used to evaluate the performance of predictions. F_β is defined as $F_\beta = (1+\beta^2)\frac{\text{Precision} \times \text{Recall}}{\beta^2 \times \text{Precision} + \text{Recall}}$,
where TP is the number of true positives, FP is the number of false positives, FN is the number of false negatives, Precision is calculated as $\frac{\text{TP}}{(\text{TP}+\text{FN})}$, and Recall is calculated as $\frac{\text{TP}}{(\text{TP}+\text{FN})}$. When $\beta > 1$, recall is emphasized over precision. When $\beta < 1$, precision is emphasized over recall. Because of the imbalanced nature of the data set with much more decoys than native proteins, we assign more weight on the small set of native proteins, with β set to 10. The F_β scores are then calculated accordingly

these 1227 proteins. To generate decoys, we fixed the composition of each of these proteins and permute its sequence by carrying out n swaps between random residues, with $n = 1, 2, 4, 8, 16, 32, 64$, and 128. The resulting decoys all have the same amino acid composition as the original native proteins, but have progressively more point mutations. We generate 1000 random sequence decoys at each swap n for each protein. We call this Decoy Set 2.

Our results show that the number of misclassified decoys decreases rapidly as the number of swaps increases. When n increases from 1 to 32, the percentage of misclassified decoys for protein of length ≤ 250 is about 30% or less. Less than 30% of the decoys of all lengths are misclassified when $n = 64$, with the rate of misclassification much smaller than 10% among those with length <350 (Fig. 4b). Only 62 decoys are misclassified among 1,227,000 decoys when $N \geq 128$ (Fig. 4b).

It is informative to examine the number of misclassified decoys and the sequence identity of the decoys with their corresponding native proteins at different protein lengths. Figure 4c shows that the percentage of misclassified decoys decreases

Table 4 Effect of the size of the preselection of dataset using Strategy 2

Preselect native proteins top (%)	Preselect decoys top	Iteration	Training Set Native/Decoy 800/36 M	F_β	Test set Native/Decoy 428/11 M	F_β
0	1	6	8/1010	0.978	25/212	0.938
2	1	5	5/1079	0.981	24/266	0.939
5	1	5	5/1038	0.981	24/247	0.939
8	1	5	5/1093	0.981	24/249	0.939
10	1	5	5/997	0.982	24/242	0.939
20	1	6	9/625	0.981	26/174	0.936
30	1	6	9/689	0.980	24/211	0.940
40	1	6	8/869	0.980	25/218	0.937
50	1	5	4/988	0.983	20/218	0.949
60	1	5	6/1039	0.980	24/280	0.938
10	1	5	5/997	0.982	24/242	0.939
10	2	5	6/1270	0.977	22/372	0.941
10	3	7	9/934	0.978	22/247	0.944
10	4	5	5/1071	0.981	24/210	0.944

Test results using Strategy 2 with different sizes of the preselected native proteins, which range from 0 to 60% while the preselected decoys are fixed as the top 1 level, and with different preselected decoys, which ranges from the top 1s to the top 4s while the preselected native proteins are fixed at 10%. Misclassifications as well as the F_β scores in two tests using different numbers of native proteins and decoys are listed (see text for details)

Table 5 Discriminating large proteins from decoys

pdb	N	n	[a]Design Decoy by KDF		[c]SwissProt Decoy by KDF		
			H	Δ_{score}	n	H	Δ_{score}
1cs0.a	1073	0	2.67	N/A	8232	2.67	2.42
1g8k.a	822	545	2.07	4.18	11,997	2.07	1.69
1gqi.a	708	1002	3.03	5.16	13,707	3.03	2.16
1kqf.a	981	93	2.19	5.17	9612	2.19	1.82
1lsh.a	954	148	1.97	4.57	10,017	1.97	2.01

Discrimination of five large proteins against sequence decoys generated by gapless threading, and against additional sequence decoys generated by threading unrelated long proteins (length from 1124 to 2459) to the structures of these five proteins. Here pdb is the PDB code of the protein structure, N is the size of protein, n is the number of decoys, H is the predicted value of the scoring function, Δ_{score} is the smallest gap of score between the native protein and its decoys. The results show that all decoys can be discriminated from natives, and the smallest score gaps between native and decoys are large

rapidly with the sequence identity to the native proteins. When decoys have a sequence identity of $\leq 60\%$ with the native protein, <10% of the decoys are misclassified, and all decoys can be discriminated against at 40% identity for proteins of length ≥ 150. For proteins of length ≤ 150, most decoys with $\leq 50\%$

sequence identity can be corrected discriminated against. These observations are consistent with current understanding of protein structures, where most proteins with $\geq 70\%$ sequence identity belong to the same family [26], and those with $\geq 30\%$ sequence identity have similar structure [63].

To examine whether misclassified decoy sequences are actually more native-like and therefore more likely to potentially adopt the correct structures than those correctly classified as nonnatives, we selected 5.5 M misclassified decoys and 4.3 M correctly classified decoys from all decoys in Decoy Set 2, and examined their energy values. We use the DFIRE energy function that was developed in [90, 91]. These decoys all have values of net DFIRE energy difference of decoys to native proteins between [0.0, 1.0] kcal/mol. Our results (Fig. 4d) show that overall, misclassified decoys have much lower average DFIRE energy values than correctly classified decoys, indicating that they are potentially more native-like than those correctly classified as decoys.

Running Time

For the simplified nonlinear fitness function derived from a rectangular kernel, the algorithm was implemented in the C language. It called Lapack [2] and used LU decomposition to solve the system of linear equations. It also called an SVD routine to determine the 2-norm of a matrix for calculating $\beta = 1.1(1/C + ||DA - \boldsymbol{e}||_2^2)$. Once matrices A and $\bar{A}$ were specified, the fitness function $H(\boldsymbol{c})$ can be derived in about 2 h and 10 min on a 2 Dual Core AMD Opteron(tm) Processors of 1800 MHz with 4 Gb memory for an A of size 8800×210 and an $\bar{A}$ of size 3680×210. The evaluation of the fitness of 14 million decoys took 2 h and 10 min using 144 CPUs of a Linux cluster [2 Dual Core AMD Opteron(tm) Processors of 1.8 GHz with 2 Gb memory for each node]. Because of the large size of the data set, the bottleneck in computation is disk IO.

Discussion

Full Nonlinear Fitness Function for Global Fitness Function of Proteins

A basic requirement for computational studies of protein design is an effective fitness or scoring function, which allows searching and identifying sequences adopting the desired structural templates. The goal of this study is to explore ways to improve the sensitivity and/or specificity of discrimination.

There are several routes toward improving empirical fitness functions. One approach is to introduce higher order interactions, where three-body or four-body interactions are explicitly incorporated in the fitness function [6, 47, 59, 62, 93]. We develop a different framework for developing empirical protein fitness functions, with the goal of simultaneous characterization of fitness landscapes of many proteins. We use a set of Gaussian kernel functions located at both native proteins and decoys as the basis set. Decoy set in this formulation is equivalent to the reference state or null model used in statistical potential. The expansion coefficients $\{\alpha_N\}, N \in \mathcal{N}$ and $\{\alpha_D\}, D \in \mathcal{D}$ of the Gaussian kernels determine the specific form of the fitness function. Since native proteins and decoys are nonredundant and are represented as unique vectors $\boldsymbol{c} \in \mathbb{R}^d$, the Gram matrix of the kernel function is full rank. Therefore, the kernel function effectively maps the protein space into a high-dimensional space in which effective discrimination with a hyperplane is easier to obtain. The optimization criterion here is not Z-score, rather we search for the hyperplane in the transformed high-dimensional space with maximal separation distance between the native protein vectors and the decoy vectors. This choice of optimality criterion is firmly rooted in a large body of studies in statistical learning theory, where expected number of errors in classification of unseen future test data is minimized probabilistically by balancing the minimization of the training error (or *empirical risk*) and the control of the capacity of specific types of functional form of the fitness function [8, 67, 80].

This approach is general and flexible, and can accommodate other protein representations, as long as the final descriptor of protein and decoy is a d-dimensional vector. In addition, different forms of nonlinear functions can be designed using different kernel functions.

Nonlinear Fitness Function Generalizes Well: Global Fitness Function Can Discriminate Dissimilar Proteins

As any other discrimination problems, the success of classification strongly depends on the training data. If the fitness function is challenged with a drastically different protein than proteins in the training set, the classification may fail. To further test how well the nonlinear fitness function performs when discriminating proteins that are dissimilar to those contained in the training set, we take five proteins that are longer than any training proteins (lengths between 46 and 688). These are obtained from the list of 1261 polypeptide chains contained in the updated Oct. 15, 2002 release of WHATIF database. The first test is to discriminate the 5 proteins from 1728 exhaustively generated design decoys using gapless threading. The second test is to discriminate these 5 proteins from exhaustively enumerated sequence decoys generated by threading 14 large protein sequences of unknown structures obtained from SwissProt database, whose sizes are between 1124 and 2459 (Table 5). This is necessary since structures of the longest chains otherwise have few or no threading decoys. Table 5 lists results of these tests, including the predicted score value and the smallest gap between the native protein and decoys. For the first test, the nonlinear design fitness functions can discriminate these 5 native proteins from all decoys. For the second test, the design fitness function can also discriminate all 5 proteins from a total of 53,565 SwissProt sequence decoys, and the smallest score gaps between native and decoys are large.

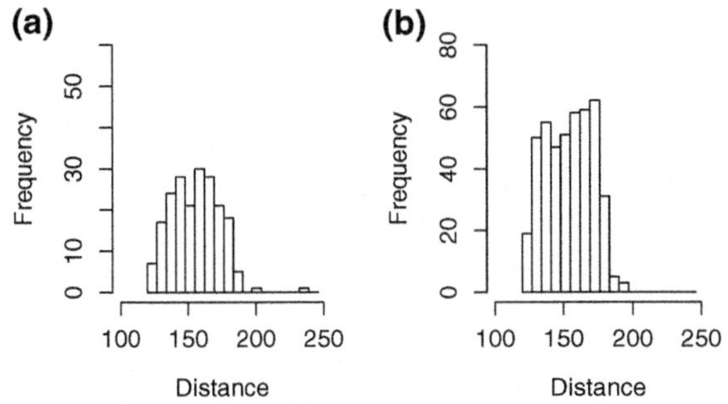

Fig. 3 The distribution of maximum distances of proteins to the set of training proteins. **a** The maximum distance for each training protein to all other 439 proteins. **b** The maximum distance for each protein in the 201 test set to all 440 training proteins. These two distributions are similar

It is infrequent for an unknown test protein to have low similarity to all reference proteins. For each protein in the 440 training set, we calculate its Euclidean distance to the other 439 proteins. The distribution of the 440 maximum distances for each training protein to all other 439 proteins is shown in Fig. 3a. We also calculate for each protein in the 201 test set its maximum distance to all training proteins (Fig. 3b). It is clear that for most of the 201 test proteins, the values of maximum distances to training proteins are similar to the values for training set proteins. The only exceptions are two proteins, ribonuclease inhibitor (`1a4y.a`) and formaldehyde ferredoxin oxidoreductase (`1b25.a`). Although they are correctly classified, the former has significant amount of unaccounted interchain contact with another protein angiogenin, and the latter has iron/sulfur clusters. It seems that the set of training proteins provide an adequate basis set for characterizing the global fitness landscape of sequence design for other proteins.

Simplified Nonlinear Fitness Function

We have also developed a simplified nonlinear kernel function for fitness landscape of protein design using a rectangular kernel and a fast Newton method. The results in a blind test are encouraging. They suggest that for a simplified task of designing simultaneously 428 proteins from a set of 11 million decoys, the search space of protein shape and sequence can be effectively parameterized with just about 3680 basis set of contact vectors. It is likely that the choice of matrix A is important. We showed that once A is carefully chosen, the overall design landscape is not overly sensitive to the specific choice of the basis set contact vectors for $\bar{A}$.

The native protein list in both training and test sets for the simplified nonlinear fitness function come from the PISCES server, which has the lowest pairwise identity (20%), finer resolution cutoff (1.6 Å), and lower R-factor cutoff (0.25). This native dataset is better than the dataset derived from the WHATIF database, which has looser constraints: pairwise sequence identity <30%, resolution cutoff <2.1 Å, and R-factor cutoff <2.1. We compared our results with classic studies of Tobi et al. [79], Bastolla et al. [4], and Miyazawa and Jernigan [58]. Although the training set and test set are different, we observed that our simplified nonlinear function detected 95% (208) native proteins from 11 million decoys and only misclassified 218 decoys as native proteins, which outperformed Tobi et al. [79] (78% correct rate for native proteins, 53,137 misclassification for decoys), Bastolla et al. [4] (71% correct rate for native proteins, 29,309 misclassification for decoys), and Miyazawa and Jernigan [58] methods (57% correct rate for native proteins, 80,716 misclassification for decoys) on much smaller blind test set of 201 native proteins and 3 million decoys.

Our final fitness landscape using rectangle kernel can correctly classify most of the native proteins, except 4 proteins (1ft5 chain A, 1gk9 chain A, 2p0s chain A, 2qud chain A) in the training set and 20 proteins in the test set. Of the 4 misclassified training proteins, all have ligand or organic molecules bound. For example, cytochrome C554 (1ft5, chain A) is a electron transport protein with 4 hemes bound, and ABC transporter (2p0s, chain A) has a Mg ion bound. Overall, among the misclassified proteins, 14 proteins contain metal ions and organic compounds. We note that the interactions between these organic compounds, metal ions, and rest of the protein are also not reflected in the protein description. In addition, 4 proteins have >20% contacts due to interchain interactions. It is likely that substantial unaccounted interactions with other protein chains, DNA, or cofactors contributed to the misclassifications. The conformations of these proteins may be different upon removal of these contacts. Altogether, 21 of the 24 misclassified proteins have explanations, and the fitness function truly failed only for 3 proteins.

The representation of protein structures will likely have important effects on the success of protein design. The approach of the reduced nonlinear function is general and applicable when alternative representations of protein structures are used, e.g., adding solvation terms, including higher order interactions.

The nonlinear fitness function and computational procedure we developed permit more accurate and rapid recognition of designed proteins which can fold into desired structures. It generalizes well and can recognize novel protein folds that are not encountered in the training process. Such an improvement in algorithms for computational protein design is essential for the success of large-scale efforts in identifying sequences for improved biochemical functions or new enzymes, so the appropriate scaffold can be constructed to which the necessary catalytical mechanism can be inserted [3]. A drawback of this nonlinear fitness function is that training is based on only native sequences of known protein structures, and therefore there is no guarantee that it can distinguish high-resolution sequences with just a few deleterious point mutations that cannot fold. Further improvement can be achieved by incorporating in training high-density homologous sequences that are known to fold into the same structural fold, as well as information on critical residues whose mutations would unfold the proteins. Protein design has had numerous successes in introducing new proteins and peptides for therapeutic applications [34], and will continue to be important for developing effective therapeutics.

Conclusion

Our findings show that there is no fitness function that can discriminate a training set of 440 native sequence from 14 million sequence decoys generated by gapless threading. The success of nonlinear fitness function in perfect discrimination of this training set proteins and its good performance in an unrelated test set of 194 proteins is encouraging. It indicates that it is now possible to characterize simultaneously the fitness landscape of many proteins, and nonlinear kernel fitness function is a general strategy for developing effective fitness function for protein sequence design.

Our study of fitness function for sequence design is a much smaller task than developing a full-fledged fitness function, because we study a restricted version of the protein design problem. We need to recognize only one sequence that folds into a known structure from other sequences already known to be part of a different protein structure, whose identity is hidden during training. However, this simplified task is challenging, because the native sequences and decoy sequences in this case are all taken from real proteins. Success in this task is a prerequisite for further development of a full-fledged universal fitness function. A full solution to the sequence design problem will need to incorporate additional sequences of structural homologs as native sequences, as well as additional decoys sequences that fold into different folds, and decoy sequences that are not proteins (e.g., all hydrophobes). Results presented in Fig. 4b provide some indication, where it was found that homologs generated by multiple random residue swaps of $\geq$80–85% overall sequence identities are likely to be correctly classified. Furthermore, an additional interesting test of our nonlinear function is to discriminate decoys generated by a method independent of the threading method used for generating training sequence decoys. For example, decoys generated by a protein design tool can be used to test how well our nonlinear scoring function can discriminate those that would fold from those that would not. This, however, requires experimental knowledge of which of these decoys indeed would fold into stable structures and which do not. It would be desirable if such data can be used at a large scale for globally all known protein folds. It is our hope that the functional form and the optimization technique introduced here will also be useful for such purposes.

We also showed that a simplified nonlinear fitness function for protein design can be obtained using a simplified nonlinear kernel function via a finite Newton method. We used a rectangular kernel with a basis set of native proteins and decoys chosen a priori. We succeeded in predicting 408 out of the 428 (95%) native proteins and misclassified only 218 out of 11 million decoys in a large blind test set. Although the test set used is different, other methods were based on relatively small blind test sets (e.g., 201 native proteins and 3 million decoys). Our result outperforms statistical linear scoring function (87 out of the 201 misclassifications, 57% correct rate) and optimized linear function (between 44 and 58 misclassifications out of the 201, 78 and 71% correct rate). The performance is also comparable with results obtained from a far more complex nonlinear fitness function with > 5000 terms (18 misclassifications, 91% correct rate). Our results further suggest that for

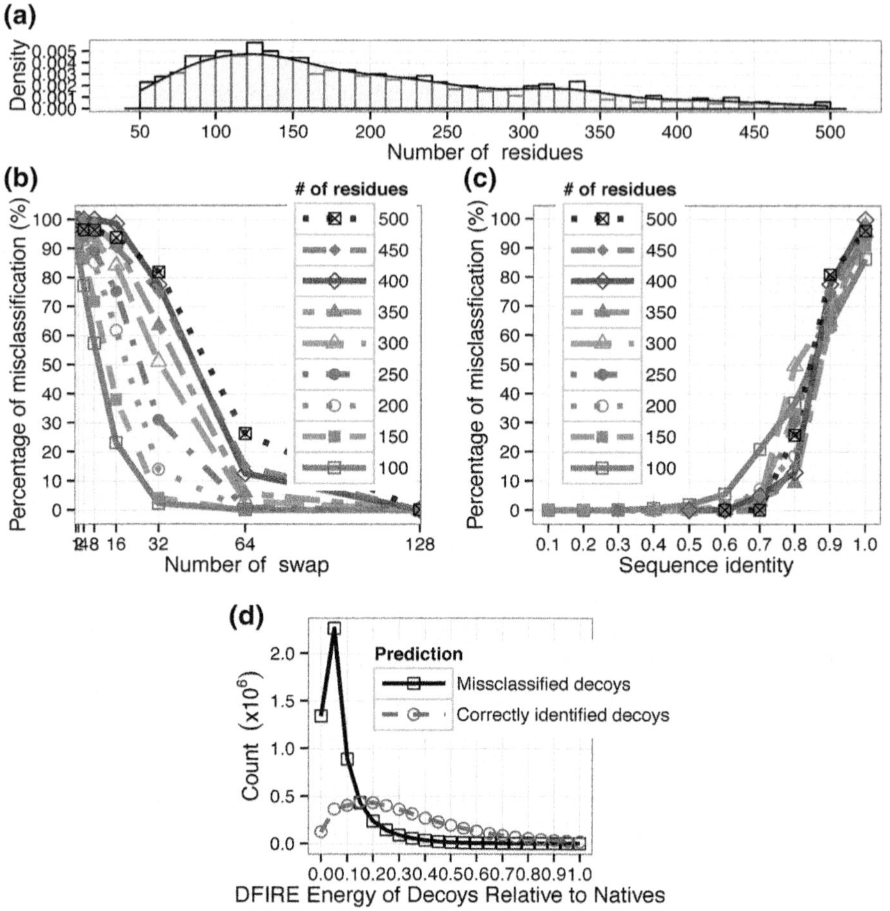

Fig. 4 Discriminating a different decoy set using the nonlinear fitness function. Sequence decoys in this set are generated by swapping residues at different positions. **a** The length distribution of the 1227 native proteins in the set; **b** The relationship between the number of swaps N and the percentage of misclassified decoys grouped by protein length binned with a width of 50 residues shown in different curves. **c** The relationship between the sequence identity binned with width 0.1 and the percentage of misclassification grouped by protein length shown in different curves. The fitness function was derived using Strategy 2, with top 50% preselected native proteins, and top 1 preselected decoys. **d** Misclassified sequence decoys have overall lower DFIRE energy values than correctly classified sequence decoys and therefore are more native-like. The x-axis is the net DFIRE energy difference of decoys to native proteins, and the y-axis is the number count of decoys at different net DFIRE energy differences. The solid black line represents decoys misclassified by our fitness function and the *dashed red line* represents decoys correctly classified by our fitness function

the task of global sequence design of 428 selected proteins, the search space of protein shape and sequence can be effectively parameterized with just about 3680 carefully chosen basis sets of native proteins and nonnative protein decoys.

In summary, we show a formulation of fitness function using a mixture of Gaussian kernels. We demonstrate that this formulation can lead to effective design scoring function that characterize fitness landscape of many proteins simultaneously, and perform well in blind independent tests. Our results suggest that this functional form different from the simple weighted sum of contact pairs can be useful for studying protein design. In addition, the approach of the rectangle kernel matrix with a finite Newton method works well in constructing fitness landscape. We also showed that the overall landscape is not overly sensitive to the specific choice of the dataset. Our approach can be generalized for any other protein representation, e.g., with descriptors for explicit hydrogen bond and higher order interactions, and our strategy of reduced kernel can be generalized to constructing other types of fitness function. Overall, constructing a universal fitness landscape that explains all major protein structural folds is a fundamental problem. In our study, an overly simplistic assumption is made, in which the only determinant of protein fitness function is the ability to fold correctly. More realistic fitness function should include functional fitness such as efficiencies in biochemical reactions. Furthermore, once a fitness function is constructed, it will be important to analyze the evolutionary landscape of proteins globally and to decipher the corresponding structural implications. It would also be useful to identify most probable transition paths among different protein folds to gain understanding on how protein structures evolve and how new folds are acquired, as well as possible timing of the emergency of such new folds.

References

1. G.A. Lazar, J.R. Desjarlais, T.M. Handel, De novo design of the hydrophobi core of ubiquitin. Protein Sci. **6**, 1167–1178 (1997)
2. E. Anderson, Z. Bai, C Bischof, *LAPACK Users' Guide*. (Society for Industrial Mathematics, 1999)
3. D. Baker, An exciting but challenging road ahead for computational enzyme design. Protein Sci. **19**(10), 1817–1819 (2010). doi:10.1002/pro.481, URLhttp://dx.doi.org/10.1002/pro.481
4. U. Bastolla, J. Farwer, E.W. Knapp, M. Vendruscolo, How to guarantee optimal stability for most representative structures in the protein data bank. Proteins **44**(2), 79–96 (2001)
5. A. Ben-Naim, Statistical potentials extracted from protein structures: are these meaningful potentials? J. Chem. Phys. **107**, 3698–3706 (1997)
6. M.R. Betancourt, D. Thirumalai, Pair potentials for protein folding: choice of reference states and sensitivity of predicted native states to variations in the interaction schemes. Protein Sci. **8**, 361–369 (1999)
7. D.N. Bolon, S.L. Mayo, Enzyme-like proteins by computational design. Proc. Natl. Acad. Sci. U.S.A. **98**(25), 14274–14279 (2001)
8. C.J.C. Burges, A tutorial on support vector machines for pattern recognition. Data Min. knowl. Disc. **2**(2), 121–167 (1998). URL http://www.kernel-machines.org/papers/Burges98.ps.gz
9. T.L. Chiu, R.A. Goldstein, Optimizing energy potentials for success in protein tertiary structure prediction. Fold Des. **3**, 223–228 (1998)

10. B.I. Dahiyat, S.L. Mayo, De novo protein design: fully automated sequence selection. Science **278**, 82–87 (1997)
11. W.F. DeGrado, C.M. Summa, V. Pavone, F. Nastri, A. Lombardi, De novo design and structural characterization of proteins and metalloproteins. Annu. Rev. Biochem. **68**, 779–819 (1999)
12. J.R. Desjarlais, T.M. Handel, De novo design of the hydrophobic cores of proteins. Protein Sci. **19**, 244–255 (1995)
13. J.M. Deutsch, T. Kurosky, New algorithm for protein design. Phys. Rev. Lett. **76**(2), 323–326 (1996)
14. R.I. Dima, J.R. Banavar, A. Maritan, Scoring functions in protein folding and design. Protein Sci. **9**, 812–819 (2000)
15. K.E. Drexler, Molecular engineering: an approach to the development of general capabilities for molecular manipulation. Proc. Natl. Acad. Sci. U.S.A. **78**, 5275–5278 (1981)
16. H. Edelsbrunner, *Algorithms in Combinatorial Geometry* (Springer, Berlin, 1987)
17. H. Edelsbrunner, The union of balls and its dual shape. Discrete Comput. Geom. **13**, 415–440 (1995)
18. E.G. Emberly, N.S. Wingreen, C.Tang, Designability of alpha-helical proteins. Proc. Natl. Acad. Sci. U.S.A. **99**(17), 11163–11168 (2002)
19. M.S. Friedrichs, P.G. Wolynes, Toward protein tertiary structure recognition by means of associative memory hamiltonians. Science **246**, 371–373 (1989)
20. G. Fung, O.L. Mangasarian, Finite Newton method for Lagrangian support vector machine classification. Neurocomputing **55**, 39–55 (2003)
21. G. Vriend, C. Sander, Quality control of protein models—directional atomic contact analysis. J. Appl. Cryst. **26**, 47–60 (1993)
22. R. Goldstein, Z.A. Luthey-Schulten, P.G. Wolynes, Protein tertiary structure recognition using optimized Hamiltonians with local interactions. Proc. Natl. Acad. Sci. U.S.A. **89**, 9029–9033 (1992)
23. M.H. Hao, H. Scheraga, Designing potential energy functions for protein folding. Curr. Opin. Struct. Biol. **9**, 184–188 (1999)
24. M.H. Hao, H.A. Scheraga, How optimization of potential functions affects protein folding. Proc. Natl. Acad. Sci. **93**(10), 4984–4989 (1996)
25. R.B. Hill, D.P. Raleigh, A. Lombardi, W.F. DeGrado, De novo design of helical bundles as models for understanding protein folding and function. Acc. Chem. Res. **33**(11), 745–754 (2000)
26. L. Holm, C. Ouzounis, C. Sander, G. Tuparev, G. Vriend, A database of protein structure families with common folding motifs. Protein Sci. (A publication of the Protein Society) **1** (12), 1691–1698 (1992)
27. C. Hu, X. Li, J. Liang, Developing optimal non-linear scoring function for protein design. Bioinformatics (Oxford, England) **20**(17), 3080–3098 (2004)
28. R.L. Jernigan, I. Bahar, Structure-derived potentials and protein simulations. Curr. Opin. Struct. Biol. **6**, 195–209 (1996)
29. L. Jiang, E.A. Althoff, F.R. Clemente, L. Doyle, D. Röthlisberger, A. Zanghellini, J.L. Gallaher, J.L. Betker, F. Tanaka, C.F. Barbas, D. Hilvert, K.N. Houk, B.L. Stoddard, D. Baker, De novo computational design of retro-aldol enzymes. Science (New York, NY) **319** (5868), 1387–1391 (2008)
30. L.A. Joachimiak, T. Kortemme, B.L. Stoddard, D. Baker, Computational design of a new hydrogen bond network and at least a 300-fold specificity switch at a protein-protein interface. J. Mol. Biol. **361**(1), 195–208 (2006)
31. T. Joachims, Making large-scale SVM learning practical, in *Advances in Kernel Methods—Support Vector Learning*, ed. by B. Scho¨lkopf, C. Burges, A. Smola (MIT Press, 1999)
32. D.T. Jones, W.R. Taylor, J.M. Thornton, A new approach to protein fold recognition. Nature **358**, 86–89 (1992)
33. N. Karmarkar, A new polynomial-time algorithm for linear programming. Combinatorica **4**, 373–395 (1984)

34. G.A. Khoury, J. Smadbeck, C.A. Kieslich, C.A. Floudas, Protein folding and de novo protein design for biotechnological applications. Trends Biotechnol. **32**(2), 99–109 (2014). doi:10.1016/j.tibtech.2013.10.008, URL http://www.sciencedirect.com/science/article/pii/S0167779913002266
35. J.M. Kleinberg, Efficient algorithms for protein sequence design and the analysis of certain evolutionary fitness landscapes. J. Comput. Biol. (A journal of computational molecular cell biology) **6**(3–4), 387–404 (1999)
36. P. Koehl, M. Levitt, De novo protein design. I. In search of stability and specificity. J. Mol. Biol. **293**, 1161–1181 (1999)
37. P. Koehl, M. Levitt, De novo protein design. II. Plasticity of protein sequence. J. Mol. Biol. **293**, 1183–1193 (1999)
38. K.K. Koretke, Z. Luthey-Schulten, P.G. Wolynes, Self-consistently optimized statistical mechanical energy functions for sequence structure alignment. Protein Sci. **5**, 1043–1059 (1996)
39. K.K. Koretke, Z. Luthey-Schulten, P.G. Wolynes, Self-consistently optimized energy functions for protein structure prediction by molecular dynamics. Proc. Natl. Acad. Sci. **95** (6), 2932–2937 (1998)
40. B. Kuhlman, D. Baker, Native protein sequences are close to optimal for their structures. Proc. Natl. Acad. Sci. U.S.A. **97**, 10383–10388 (2000)
41. B. Kuhlman, G. Dantas, G.C. Ireton, G. Varani, B.L. Stoddard, D. Baker, Design of a novel globular protein fold with atomic-level accuracy. Science **302**, 1364–1368 (2003)
42. G.A. Lazar, W. Dang, S. Karki, O. Vafa, J.S. Peng, L. Hyun, C. Chan, H.S. Chung, A. Eivazi, S.C. Yoder, J. Vielmetter, D.F. Carmichael, R.J. Hayes, B.I. Dahiyat, Engineered antibody Fc variants with enhanced effector function. Proc. Natl. Acad. Sci. U.S.A. **103**(11), 4005–4010 (2006)
43. Y.J. Lee, O.L. Mangasarian, RSVM: Reduced support vector machines, in *Proceedings of the First SIAM International Conference on Data Mining* (2001), pp. 1–17
44. C.M.R. Lemer, M.J. Rooman, S.J. Wodak, Protein-structure prediction by threading methods —evaluation of current techniques. Proteins **23**, 337–355 (1995)
45. H. Li, R. Helling, C. Tang, N. Wingreen, Emergence of preferred structures in a simple model of protein folding. Science **273**, 666–669 (1996)
46. X. Li, J. Liang, Cooperativity and anti-cooperativity of three-body interactions in proteins. J. Phys. Chem. B (In review) (2004)
47. X. Li, C. Hu, J. Liang, Simplicial edge representation of protein structures and alpha contact potential with confidence measure. Proteins **53**, 792–805 (2003)
48. J. Liang, H. Edelsbrunner, P. Fu, P.V. Sudhakar, S. Subramaniam, Analytical shape computing of macromolecules I: Molecular area and volume through alpha-shape. Proteins **33**, 1–17 (1998)
49. H. Lu, J. Skolnick, A distance-dependent atomic knowledge-based potential for improved protein structure selection. Proteins **44**, 223–232 (2001)
50. V.N. Maiorov, G.M. Crippen, Contact potential that recognizes the correct folding of globular proteins. J. Mol. Biol. **227**, 876–888 (1992)
51. O.L. Mangasarian, *Nonlinear Programming* (Society for Industrial Mathematics, 1994)
52. J. Meller, M. Wagner, R. Elber, Maximum feasibility guideline in the design and analysis of protein folding potentials. J. Comput. Chem. **23**, 111–118 (2002)
53. C.S. Mészáros, Fast Cholesky factorization for interior point methods of linear programming. Comput. Math. Appl. **31**, 49–51 (1996)
54. C. Micheletti, F. Seno, J.R. Banavar, A. Maritan, Learning effective amino acid interactions through iterative stochastic techniques. Proteins **42**(3), 422–431 (2001)
55. L.A. Mirny, E.I. Shakhnovich, How to derive a protein folding potential? A new approach to an old problem. J. Mol. Biol. **264**, 1164–1179 (1996)
56. S. Miyazawa, R. Jernigan, Estimation of effective interresidue contact energies from protein crystal structures: quasi-chemical approximation. Macromolecules **18**, 534–552 (1985)

57. S. Miyazawa, R. Jernigan, Residue-residue potentials with a favorable contact pair term and an unfavorable high packing density term. J. Mol. Biol. **256**, 623–644 (1996). URL citeseer.nj.nec.com/388482.html
58. S. Miyazawa, R.L. Jernigan, Residue-residue potentials with a favorable contact pair term and an unfavorable high packing density term, for simulation and threading. J. Mol. Biol. **256**(3), 623–644 (1996)
59. P.J. Munson, R.K. Singh, Statistical significance of hierarchical multi-body potential based on Delaunay tessellation and their application in sequence-structure alignment. Protein Sci. **6**, 1467–1481 (1997)
60. J. Nocedal, S.J. Wright, *Numerical Optimization* (Springer, 1999)
61. C. Pabo, Designing proteins and peptides. Nature **301**, 200 (1983)
62. A. Rossi, C. Micheletti, F. Seno, A. Maritan, A self-consistent knowledge-based approach to protein design. Biophys. J. **80**(1), 480–490 (2001)
63. B. Rost, Twilight zone of protein sequence alignments. Protein Eng. Des. Sel.: PEDS **12**(2), 85–94 (1999)
64. D. Röthlisberger, O. Khersonsky, A.M. Wollacott, L. Jiang, J. DeChancie, J. Betker, J.L. Gallaher, E.A. Althoff, A. Zanghellini, O. Dym, S. Albeck, K.N. Houk, D.S. Tawfik, D. Baker, Kemp elimination catalysts by computational enzyme design. Nature **453**(7192), 190–195 (2008)
65. R. Samudrala, J. Moult, An all-atom distance-dependent conditional probability discriminatory function for protein structure prediction. J. Mol. Biol. **275**, 895–916 (1998)
66. B. Schölkopf, *Learning with Kernels: Support Vector Machines, Regularization, Optimization, and Beyond* (The MIT Press, 2002)
67. B. Schölkopf, A.J. Smola, *Learning with Kernels: Support Vector Machines, Regularization, Optimization, and Beyond* (The MIT Press, Cambridge, 2002)
68. E.I. Shakhnovich, Protein design: a perspective from simple tractable models. Fold Des. **3**, R45–R58 (1998)
69. E.I. Shakhnovich, A.M. Gutin, Engineering of stable and fast-folding sequences of model proteins. Proc. Natl. Acad. Sci. U.S.A. **90**, 7195–7199 (1993)
70. J.M. Shifman, M.H. Choi, S. Mihalas, S.L. Mayo, M.B. Kennedy, Ca2+/calmodulin-dependent protein kinase II (CaMKII) is activated by calmodulin with two bound calciums. Proc. Natl. Acad. Sci. U.S.A. **103**(38), 13968–13973 (2006)
71. J.B. Siegel, A. Zanghellini, H.M. Lovick, G. Kiss, A.R. Lambert, J.L. St Clair, J.L. Gallaher, D. Hilvert, M.H. Gelb, B.L. Stoddard, K.N. Houk, F.E. Michael, D. Baker, Computational design of an enzyme catalyst for a stereoselective bi- molecular Diels-Alder reaction. Science (New York, NY) **329**(5989), 309–313
72. K.T. Simons, I. Ruczinski, C. Kooperberg, B. Fox, C. Bystroff, D. Baker, Improved recognition of native-like protein structures using a combination of sequence-dependent and sequence-independent features of proteins. Proteins **34**, 82–95 (1999)
73. M.J. Sippl, Knowledge-based potentials for proteins. Curr. Opin. Struct. Biol. **5**(2), 229–235 (1995)
74. A.M. Slovic, H. Kono, J.D. Lear, J.G. Saven, W.F. DeGrado, From the Cover: Computational design of water-soluble analogues of the potassium channel KcsA. Proc. Natl. Acad. Sci. U.S. A. **101**(7), 1828–1833 (2004)
75. S. Tanaka, H.A. Scheraga, Medium- and long-range interaction parameters between amino acids for predicting three-dimensional structures of proteins. Macromolecules **9**, 945–950 (1976)
76. P.D. Thomas, K.A. Dill, An iterative method for extracting energy-like quantities from protein structures. Proc. Natl. Acad. Sci. U.S.A. **93**, 11628–11633 (1996)
77. P.D. Thomas, K.A. Dill, Statistical potentials extracted from protein structures: how accurate are they? J. Mol. Biol. **257**, 457–469 (1996)
78. D. Tobi, G. Shafran, N. Linial, R. Elber, On the design and analysis of protein folding potentials. Proteins **40**, 71–85 (2000)

79. D. Tobi, G. Shafran, N. Linial, R. Elber, On the design and analysis of protein folding potentials. Proteins **40**(1), 71–85 (2000)
80. V. Vapnik, *The Nature of Statistical Learning Theory* (Springer, New York, 1995)
81. V. Vapnik, *The Nature of Statistical Learning Theory* (Information Science and Statistics), 2nd edn. (Springer, 1999)
82. V. Vapnik, A. Chervonenkis, A note on one class of perceptrons. Autom. Remote Control **25** (1964)
83. V.N. Vapnik, A.J. Chervonenkis, *Theory of Pattern Recognition* [in Russian] (Nauka, Moscow, 1974) [German Translation: W. Wapnik, A. Tscherwonenkis, *Theorie der Zeichenerkennung* (Akademie–Verlag, Berlin, 1979)]
84. M. Vendruscolo, E. Domanyi, Pairwise contact potentials are unsuitable for protein folding. J. Chem. Phys. **109**(11), 101–108 (1998)
85. M. Vendruscolo, R. Najmanovich, E. Domany, Can a pairwise contact potential stabilize native protein folds against decoys obtained by threading? Proteins **38**, 134–148 (2000)
86. M. Vendruscolo, R. Najmanovich, E. Domany, Can a pairwise contact potential stabilize native protein folds against decoys obtained by threading? Proteins: Struct. Funct. Genet. **38**, 134–148 (2000)
87. G. Wang, R.L. Dunbrack, PISCES: a protein sequence culling server. Bioinformatics (Oxford, England) **19**(12), 1589–1591 (2003)
88. L. Wernisch, S. Hery, S.J. Wodak, Automatic protein design with all atom force-fields by exact and heuristic optimization. J. Mol. Biol. **301**, 713–736 (2000)
89. S.J. Wodak, M.J. Rooman, Generating and testing protein folds. Curr. Opin. Struct. Biol. **3**, 247–259 (1993)
90. Y. Yang, Y. Zhou, Ab initio folding of terminal segments with secondary structures reveals the fine difference between two closely related all-atom statistical energy functions. Protein Sci. **17**(7), 1212–1219 (2008)
91. Y. Yang, Y. Zhou, Specific interactions for ab initio folding of protein terminal regions with secondary structures. Proteins **72**(2), 793–803 (2008)
92. K. Yue, K.A. Dill, Inverse protein folding problem: designing polymer sequences. Proc. Natl. Acad. Sci. U.S.A. **89**, 4163–4167 (1992)
93. W. Zheng, S.J. Cho, I.I. Vaisman, A. Tropsha, A new approach to protein fold recognition based on Delaunay tessellation of protein structure, in *Pacific Symposium on Biocomputing'97*, ed. by R. Altman, A. Dunker, L. Hunter, T. Klein (World Scientific, Singapore, 1997), pp. 486–497

Computational Methods for Mass Spectrometry Imaging: Challenges, Progress, and Opportunities

Chanchala D. Kaddi and May D. Wang

Abstract Mass spectrometry imaging (MSI) is a rapidly growing field of research, with applications in proteomics, lipidomics, and metabolomics. The benefit of MSI is its capacity to measure spatially resolved molecular information. Computational methods are important to extracting information from MSI data for basic and translational research. In this chapter, we examine current and emerging methods for analysis of MSI data, and highlight associated challenges and opportunities in computational research for MSI.

Introduction

Mass spectrometry imaging (MSI) is a large-scale experimental technique that can yield spatially resolved information about the molecular composition of a biological sample. MSI datasets are generated by acquiring the complete mass spectrum at multiple points across the sample surface, yielding a three-dimensional (x, y: spatial, e.g., tissue, and z: spectral or m/z) dataset as shown in Fig. 1.

The MSI dataset includes valuable information which is not obtainable through similar analyses using immunohistochemistry staining or non-imaging mass spectrometry. In traditional histological analysis, tissue is typically stained for a small number of molecular targets; in contrast, MSI is capable of simultaneously tracking thousands of m/z (mass-to-charge ratio) values. Depending on the MSI acquisition modality, each m/z value can be interpreted as a molecule or molecular fragment. Additionally, staining can only identify known molecular targets, while the large-scale data acquired by MSI enables discovery of sample components (and hence, potential biomarkers). Compared to mass spectrometry alone, MSI preserves

---

C.D. Kaddi · M.D. Wang (✉)
Georgia Institute of Technology and Emory University, Atlanta, Georgia, USA
e-mail: maywang@bme.gatech.edu

D. Xu et al. (eds.), *Health Informatics Data Analysis*, Health Information Science,
DOI 10.1007/978-3-319-44981-4_2

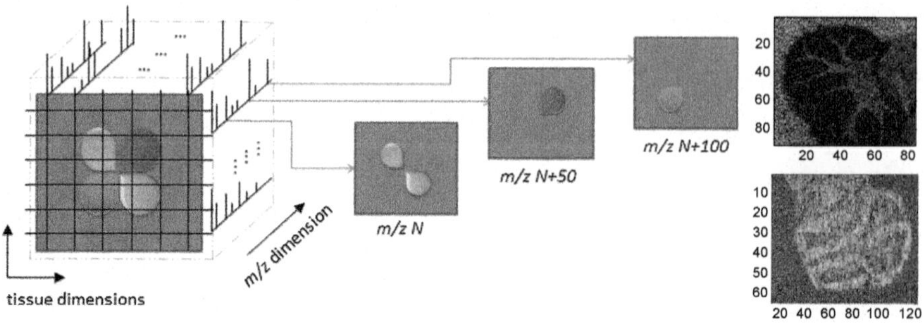

Fig. 1 (*Left*) Three-dimensional structure of MSI data. (*Right*) False-color visualizations of multiple *m/z* values from MSI datasets of mouse models of Tay-Sachs/Sandhoff disease

the sample's spatial and morphological information. Thus, spectra corresponding to different regions of organs, or to tumor, marginal, or normal sections of biopsies, can be differentiated, enabling more detailed and target-specific analysis. Due to these benefits, MSI is emerging as a popular experimental technique in proteomics [1], lipidomics [2], and metabolomics [3] research.

Because MSI is spatially resolved, it is particularly relevant for research into diseases which have spatially localized characteristics—particularly cancer. Recent MSI studies have investigated cancers of the head and neck [4], brain [5], breast [6], renal [7], stomach [8], prostate [9], colon [10], pancreas [11], and bladder [12]. Other recent MSI studies have targeted diseases including Tay-Sachs/Sandhoff disease [13], Behçet disease [14], Parkinson's disease [15, 16], Alzheimer's disease [17], Duchenne muscular dystrophy [18, 19], Fabry disease [20], atherosclerosis [21], and stroke and ischemic injury [22–24]. In addition, MSI has been used to study bio-implant interfaces [25, 26] and drug distribution within tissues [27–32].

The spectral dimension of MSI data can be very large (e.g., tens of thousands of *m/z* values), making computational analysis essential to interpretation. It is critical to identify and to develop effective analytical methods for large-scale data mining, pattern recognition, and exploration. This chapter begins by discussing several key open challenges in MSI research. Next, the current state-of-the-art in MSI analysis will be described, including techniques such as principal component analysis, clustering, and classification. Additionally, several emerging methods for MSI analysis, such as non-negative matrix factorization, will be introduced. All methods are discussed in the context of recent MSI studies which apply them, spanning several MSI modalities, such as Matrix-Assisted Laser Desorption/Ionization (MALDI)-MSI and Desorption Electrospray Ionization (DESI)-MSI. Finally, a case study in applying unsupervised analysis methods for pattern detection in MSI will be provided.

Challenges

As will be described in greater detail in the following sections, much progress has been made in identifying and developing analytical methods for pattern detection in MSI data. Research on this topic is still highly active. In particular, we highlight three new areas of interest in computational MSI research.

Challenge 1: Integration of MSI Data with Complementary Imaging Modalities

An emerging area of interest in MSI data analysis is the integration of MSI data to other images, such as those acquired via different MSI modalities or non-MSI data types. One example is the combination of MALDI-MSI data with magnetic resonance images [33]; another is integration of DESI-MSI data with histology images [34]. This type of integration harnesses the different strengths of the data types—for example, MRI imaging yields much higher resolution data than MSI, but does not measure molecular information like MSI. Similar reasoning is behind efforts to combine different MSI modalities: PCA and CCA have also been used to link low-mass SIMS-MSI data with high-mass MALDI-MSI data from the same brain tissue sample [35]. Because of the inherent differences in imaging modalities, several important computational challenges are in the image processing domain: for example, image alignment algorithms used to 'stich together' multiple MALDI-MSI images obtained from a large sample [36], and registration algorithms to map consecutive optical images in order to construct MSI datasets for three-dimensional samples [37].

Challenge 2: Movement Toward MSI from Three-Dimensional Samples

The movement toward MSI analysis of three-dimensional samples, as just mentioned, is an important development in the field. The studies described thus far have all implemented MSI on two-dimensional samples, e.g., very thin slices of tissue. If multiple spatially consecutive slices are taken from an organ or tumor, *m/z* images from multiple MSI datacubes can be stitched together to track the spatial distribution and expression of an *m/z* value through the original three-dimensional sample. We refer readers to the recent publication [4–32] for an example of this technique and discussion of the computational challenges associated with MSI for three-dimensional data.

Challenge 3: Reproducibility, Data Standardization, and Community Resources

MSI also provides a rich opportunity for biomarker identification. However, reproducibility of results is a major challenge. Many MSI studies consider a small number of samples, making it difficult to generalize the suitability of the analytical methods used. It would be valuable to examine alternative analytical pipelines for MSI data in a systematic manner on a variety of different MSI datasets, similar to the MAQC-II study conducted for microarray-based predictive modeling [38]. However, in addition to the obstacle of scale, such efforts are hindered because unlike microarray and RNA sequencing data, MSI data is not readily shared in public repositories. The development of community resources and infrastructure—as well as standards for quality control and transparency like Minimum Information protocols (http://mibbi.sourceforge.net/)—would facilitate this process. The recent release of mzML [39], a standardized data format for mass spectrometry, and the PRIDE proteomics data repository from the European Bioinformatics Institute for MS/MS proteomics datasets (http://www.ebi.ac.uk/pride/archive/), are therefore encouraging developments.

Current Techniques in MSI Analysis

Analytical methods for data and knowledge mining are divided into two main classes: supervised and unsupervised learning. In supervised learning, a predictive model is constructed from annotated training data, such that when a new sample is provided, the model can correctly predict the annotation of the sample. Supervised methods are further divided into two main categories: classification, in which the predicted annotation is a group label (e.g., "healthy" or "diseased"), or regression, in which it is a numerical value. For example, classification models have been developed using MALDI-MSI data to distinguish HER2 positive and HER2-negative tissues [6]; to distinguish cancerous and non-cancerous prostate tissue [40]; and to classify breast cancer sample regions as necrotic, viable/active tumor, or tumor interface region, while distinguishing them from embedding gelatin and glass or holes, using SIMS-MSI data [41]. In contrast, unsupervised learning requires no annotation or prior knowledge of the data structure. Unsupervised methods are used for exploration of the data and the identification of potential patterns; the results of these analyses can be a precursor to supervised analysis. Common unsupervised methods include dimensionality reduction and clustering. The remainder of this section will introduce unsupervised methods for MSI data analysis.

A. *Dimensionality reduction*

Principal component analysis

Principal component analysis (PCA) is currently one of the most popular techniques for exploratory data analysis in MSI. The utility of PCA is in its ability to highlight different spatial patterns present in the data, and the *m/z* values which contribute to them. Given a data matrix X of dimensions $M \times N$ (i.e., M mass spectra each containing N *m/z* values), PCA performs a linear transformation that projects the data into a different, potentially more meaningful, spectral coordinate space. The axes directions in this transformed space are defined by a set of orthogonal M-dimensional basis vectors (the principal components), and are related to the variance in X. The first principal component is the direction in which the variance of the data is maximized, and can be interpreted as the most prominent pattern in the data. The second principal component corresponds to the direction of the second highest variance in the data, and so on. After performing PCA, the p principal components which contribute to the majority of the variance in the dataset are retained. Since p is typically chosen to be $\ll N$, PCA produces a dimensionally reduced $M \times p$ dataset.

Non-negative matrix factorization

Non-negative matrix factorization (NMF) is another method that can be used for dimensionality reduction. In NMF, a data matrix X of dimensions ($M \times N$) is factored into two matrices W ($M \times k$) and H ($k \times N$), such that $X \approx WH$. This is done iteratively by minimizing the residual $\|X - WH\|^2$ such that $W, H \geq 0$ [42]. The user-selected parameter k is the number of components into which the data is separated. The matrix W is a set of basis vectors describing the *m/z* values which comprise each component. The k columns of matrix W can be interpreted as groupings of *m/z* values corresponding to prominent spatial patterns in the data. NMF has been assessed on both MALDI and DESI-MSI data [43–46]. A web-based tool, omniSpect, is also available for performing NMF on MSI data [47].

Other Dimensionality Reduction Methods

Independent component analysis (ICA) separates a mixture into components, based on the assumption that the mixture is a linear combination of statistically independent components with non-Gaussian distributions [48]. ICA has been assessed for studying intratumor heterogeneity via MSI data [46] and compared with PCA and NMF on MALDI-MSI data [43]. Like PCA, ICA presents an obstacle in terms of interpretation of the components, which can be negative. In canonical correlation analysis (CCA), two datasets, each with different dimensions, may be projected onto the feature space of the other such that the project data has maximum correlation [49]. CCA has also been used to correlate low-mass SIMS-MSI data with high-mass MALDI-MSI data, which improved image contrast and interpretation of the data [35]. Parallel factor analysis (PARAFAC) is another method often used in chemometrics for decomposing high-dimensional data; it has been tested on SIMS and LDI MSI datasets [50].

B. *Clustering*

Clustering is an unsupervised method for data analysis in which a sample (e.g., an individual mass spectrum) is allocated into a specific group (e.g., a cluster) based on a quantitative measure of the similarity. Clustering can reveal potentially meaningful structures and patterns within the data. For example, using MALDI-MSI data of gastric cancer, hierarchical clustering was used both to cluster spectra within a single tissue dataset, and also to cluster tumors from different patients [51]. LA-ICP-MSI data of rat brain tissue was analyzed by *k*-means clustering, revealing meaningful patterns in which clusters corresponded to known anatomical features [52]. Similarly, high dimensional discriminant clustering (HDDC) was applied to MALDI-MSI data of a rat brain and an intestinal-invading neuroendocrine tumor to find spectral clusters corresponding to morphological structures [53]. This type of mapping, termed spatial segmentation, is discussed further in a recent review [54]. While numerous algorithms exist for separating data into clusters [55], two of the most commonly applied methods are hierarchical and *k*-means approaches. In hierarchical clustering, a dataset *X* is separated into different levels of clusters, culminating in a dendrogram or cluster tree. The terminal leaf nodes each correspond to a single sample. In *k*-means, the dataset is separated into a predefined number *k* of clusters.

C. *Spatial similarity*

While similarity measures (in the spectral dimension) are utilized in clustering, another independent application of similarity measures is to identify *m/z* images with similar expression patterns in the spatial dimension. Figure 2 shows an

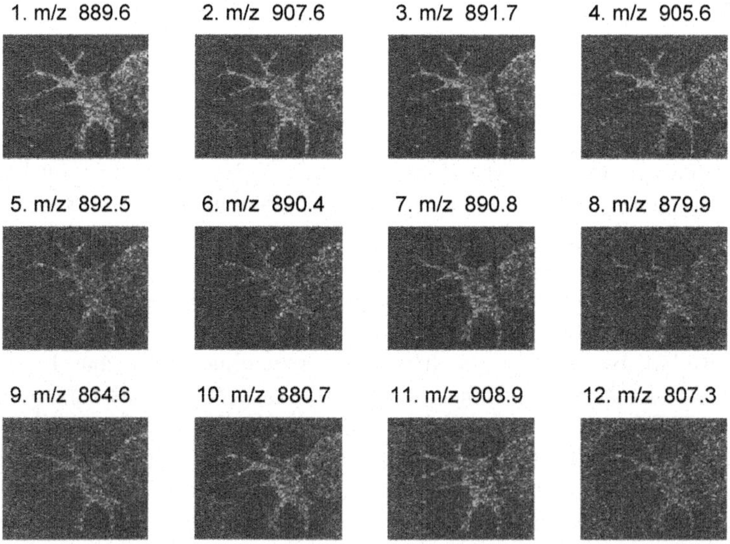

Fig. 2 Highly similar *m/z* images, as identified by the multivariate hypergeometric similarity measure [58]

example of similarity measure-based retrieval of *m/z* images with a spatial pattern similar to that of the *m/z* value of interest. For MALDI-MSI data, the similarity of *m/z* images to each other has been assessed using Pearson correlation [56] and using similarity measures based on the hypergeometric and multivariate hypergeometric distributions [57, 58]. Multivariate least-squares-based query has also been applied for this task [59].

Case Study

The following case study compares and contrasts PCA and NMF for finding potentially relevant patterns in MSI data. The data used in this example is MALDI-MSI, from a mouse model of Tay-Sachs disease [17].

Example 1 A step-by-step implementation of PCA in MATLAB R2014a to find patterns in MSI data.

Step 1: After loading the three-dimensional MSI dataset into MATLAB, restructure it into a two-dimensional matrix. Here, a and b are the spatial dimensions (i.e., the number of pixels in the horizontal and vertical directions), and c is the spectral dimension (i.e., the number of *m/z* values). In the restructured $M \times N$ matrix, M is the number of spectra ($M = a \times b$) and $N = c$.

```
[a,b,c] = size(data);
reshaped = double(reshape(data,a*b,c));
[M,N] = size(reshaped);
```

Step 2: Mean-center the data by subtracting the mean in the spectral dimension (i.e., the average of each *m/z* value).

```
mean_reshaped = mean(reshaped,1);
reshaped = reshaped − repmat(mean_reshaped,M,1);
```

Step 3: Calculate the covariance matrix and the find its eigenvectors and eigenvalues; these are the principal components and their weights. Sort the principal components in order of descending eigenvalue magnitude.

```
covariance_matrix = (1 /(N−1)) * (reshaped' *
reshaped);
[PC,V] = eig(covariance_matrix);
V = diag(V);
[~, indices] = sort(−1*V);
V = V(indices); PC = PC(:,indices);
```

Step 4: Retain the top 3 principal components, and project the original data onto these components. The reduced dataset will have dimensions $a \times b \times 3$.

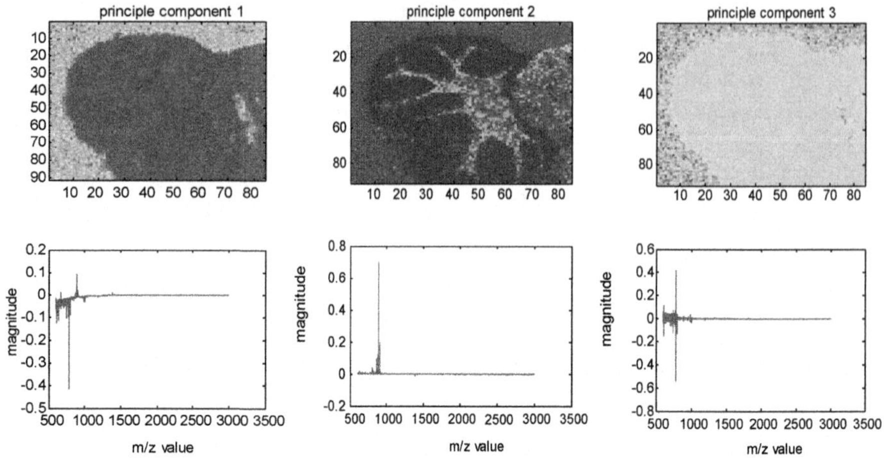

Fig. 3 The first three principal components reveal different spatial patterns and associated *m/z* values in an MSI dataset (false-color visualization)

```
projected = reshaped * PC(:,1:3);
PCA_datacube = reshape(projected,a,b,3);
```

The plots in Fig. 3 show the three images in the PCA-processed datacube (top row), and the principal components themselves (bottom row). PCA reveals different structures within the data—primarily the tissue versus non-tissue regions in PC 1 and PC 3, and cerebellum structure in PC 2.

Example 2 A step-by-step implementation of NMF in MATLAB to find patterns in MSI data.

For brevity, the MATLAB function 'nnmf' is used to perform the analysis in this example.

Step 1: Load and reshape the datacube into an $M \times N$ matrix as before.

```
[a,b,c] = size(data);
reshaped = double(reshape(data,a*b,c));
```

Step 2: Define k, the number of components, and perform NMF:

```
k = 3;
[w,h] = nnmf(reshaped,k);
NMF_datacube = reshape(w,a,b,k);
```

Figure 4 shows the three images in the NMF-processed data, i.e., the component matrix w (top row), and the corresponding row of the weight matrix h (bottom row). Similar to PCA, NMF reveals tissue versus. non-tissue patterns in the Factor 1 and Factor 2, and the cerebellum structure in Factor 3. Factor 2 also contains some information on the tissue interior. Unlike in PCA, the numerical labeling of NMF

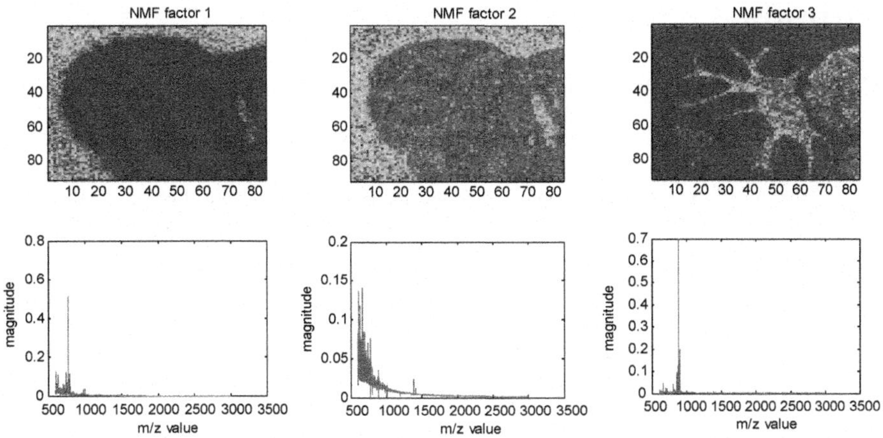

Fig. 4 The first three NMF factors reveal different spatial patterns and associated *m/*values in an MSI dataset (false-color visualization)

factors is arbitrary. Additionally, as its name indicates, the spectral profiles found by NMF are constrained to be non-negative. This is a useful property for MSI, since the data is also non-negative. In contrast, the negative values in the PCA-generated components in Fig. 3 can be difficult to interpret in terms of biology or chemistry.

Conclusion

MSI is a rapidly developing area of research with exciting implications for our understanding of numerous biological processes and diseases. In this chapter, we have described the key role of computational analyses in extracting meaningful information from MSI data. State-of-the-art techniques were described and a case study was examined. Finally, we have highlighted several challenges for computational research for MSI. In conclusion, there exist numerous opportunities for researchers to become involved in the development of computational methods and tools for MSI. Ongoing research into more effective and informative analytical techniques will help to harness the power of MSI for accelerating both basic and translational research.

Acknowledgements We thank Dr. M. Cameron Sullards and Dr. Yanfeng Chen for sharing the MSI data used in the examples in this review. This research is supported by NIH grant **U01** CA151802, Georgia Cancer Coalition distinguished cancer scholar award to Professor May D. Wang, and NSF graduate fellowship to Ms. Chanchala Kaddi.

References

1. L. MacAleese, J. Stauber, R.M.A. Heeren, Perspectives for imaging mass spectrometry in the proteomics landscape. Proteomics **9**, 819–834 (2009)
2. N. Goto-Inoue, T. Hayasaka, N. Zaima, M. Setou, Imaging mass spectrometry for lipidomics. Biochimica Et Biophysica Acta-Molecular and Cell Biology of Lipids **1811**, 961–969 (2011)
3. Y. Sugiura, M. Setou, Imaging mass spectrometry for visualization of drug and endogenous metabolite distribution: toward in situ pharmacometabolomes. J. Neuroimmune Pharmacol. **5**, 31–43 (2010)
4. S.A. Patel, A. Barnes, N. Loftus, R. Martin, P. Sloan, N. Thakker et al., Imaging mass spectrometry using chemical inkjet printing reveals differential protein expression in human oral squamous cell carcinoma. Analyst **134**, 301–307 (2009)
5. N.Y.R. Agar, J.G. Malcolm, V. Mohan, H.W. Yang, M.D. Johnson, A. Tannenbaum et al., Imaging of meningioma progression by matrix-assisted laser desorption ionization time-of-flight mass spectrometry. Anal. Chem. **82**, 2621–2625 (2010)
6. S. Rauser, C. Marquardt, B. Balluff, S.O. Deininger, C. Albers, E. Belau et al., Classification of HER2 receptor status in breast cancer tissues by MALDI imaging mass spectrometry. J. Proteome Res. **9**, 1854–1863 (2010)
7. S.R. Oppenheimer, D.M. Mi, M.E. Sanders, R.M. Caprioli, Molecular analysis of tumor margins by MALDI mass spectrometry in renal carcinoma. J. Proteome Res. **9**, 2182–2190 (2010)
8. Y. Morita, K. Ikegami, N. Goto-Inoue, T. Hayasaka, N. Zaima, H. Tanaka et al., Imaging mass spectrometry of gastric carcinoma in formalin-fixed paraffin-embedded tissue microarray. Cancer Sci. **101**, 267–273 (2010)
9. L.H. Cazares, D. Troyer, S. Mendrinos, R.A. Lance, J.O. Nyalwidhe, H.A. Beydoun et al., Imaging mass spectrometry of a specific fragment of mitogen-activated protein kinase/extracellular signal-regulated kinase kinase kinase 2 discriminates cancer from uninvolved prostate tissue. Clin. Cancer Res. **15**, 5541–5551 (2009)
10. J.W. Park, H.K. Shon, B.C. Yoo, I.H. Kim, D.W. Moon, T.G. Lee, Differentiation between human normal colon mucosa and colon cancer tissue using ToF-SIMS imaging technique and principal component analysis. Appl. Surf. Sci. **255**, 1119–1122 (2008)
11. M.C. Djidja, E. Claude, M.F. Snel, P. Scriven, S. Francese, V. Carolan et al., MALDI-Ion mobility separation-mass spectrometry imaging of glucose-regulated protein 78 kDa (Grp78) in human formalin-fixed, paraffin-embedded pancreatic adenocarcinoma tissue sections. J. Proteome Res. **8**, 4876–4884 (2009)
12. A.L. Dill, D.R. Ifa, N.E. Manicke, A.B. Costa, J.A. Ramos-Vara, D.W. Knapp et al., Lipid profiles of canine invasive transitional cell carcinoma of the urinary bladder and adjacent normal tissue by desorption electrospray ionization imaging mass spectrometry. Anal. Chem. **81**, 8758–8764 (2009)
13. Y.F. Chen, J. Allegood, Y. Liu, E. Wang, B. Cachon-Gonzalez, T.M. Cox et al., Imaging MALDI mass spectrometry using an oscillating capillary nebulizer matrix coating system and its application to analysis of lipids in brain from a mouse model of Tay-Sachs/Sandhoff disease. Anal. Chem. **80**, 2780–2788 (2008)
14. M. Aranyosiova, M. Michalka, M. Kopani, B. Rychly, J. Jakubovsky, D. Velic, Microscopy and chemical imaging of Behcet brain tissue. Appl. Surf. Sci. **255**, 1584–1587 (2008)
15. D. Hare, B. Reedy, R. Grimm, S. Wilkins, I. Volitakis, J.L. George et al., Quantitative elemental bio-imaging of Mn, Fe, Cu and Zn in 6-hydroxydopamine induced Parkinsonism mouse models. Metallomics **1**, 53–58 (2009)
16. K. Skold, M. Svensson, A. Nilsson, X.Q. Zhang, K. Nydahl, R.M. Caprioli et al., Decreased striatal levels of PEP-19 following MPTP lesion in the mouse. J. Proteome Res. **5**, 262–269 (2006)

17. R.W. Hutchinson, A.G. Cox, C.W. McLeod, P.S. Marshall, A. Harper, E.L. Dawson et al., Imaging and spatial distribution of beta-amyloid peptide and metal ions in Alzheimer's plaques by laser ablation-inductively coupled plasma-mass spectrometry. Anal. Biochem. **346**, 225–233 (2005)
18. N. Tahallah, A. Brunelle, S. De La Porte, O. Laprevote, Lipid mapping in human dystrophic muscle by cluster-time-of-flight secondary ion mass spectrometry imaging. J. Lipid Res. **49**, 438–454 (2008)
19. D. Touboul, A. Brunelle, F. Halgand, S. De La Porte, O. Laprevote, Lipid imaging by gold cluster time-of-flight secondary ion mass spectrometry: application to Duchenne muscular dystrophy. J. Lipid Res. **46**, 1388–1395 (2005)
20. D. Touboul, S. Roy, D.P. Germain, P. Chaminade, A. Brunelle, O. Laprevote, MALDI-TOF and cluster-TOF-SIMS imaging of Fabry disease biomarkers. Int. J. Mass Spectrom. **260**, 158–165 (2007)
21. N.E. Manicke, M. Nefliu, C. Wu, J.W. Woods, V. Reiser, R.C. Hendrickson et al., Imaging of lipids in atheroma by desorption electrospray ionization mass spectrometry. Anal. Chem. **18**, 8702–8707 (2009)
22. J.H. Kim, B.J. Ahn, J.H. Park, H.K. Shon, Y.S. Yu, D.W. Moon et al., Label-free calcium imaging in ischemic retinal tissue by TOF-SIMS. Biophys. J. **94**, 4095–4102 (2008)
23. S. Koizumi, S. Yamamoto, T. Hayasaka, Y. Konishi, M. Yamaguchi-Okada, N. Goto-Inoue et al., Imaging mass spectrometry revealed the production of lyso-phosphatidylcholine in the injured ischemic rat brain. Neuroscience **168**, 219–225 (2010)
24. S.X. Jiang, S. Whitehead, A. Aylsworth, J. Slinn, B. Zurakowski, K. Chan et al., Neuropilin 1 directly interacts with Fer Kinase to mediate Semaphorin 3A-induced death of cortical neurons. J. Biol. Chem. **285**, 9908–9918 (2010)
25. C. Eriksson, K. Borner, H. Nygren, K. Ohlson, U. Bexell, N. Billerdahl, et al., Studies by imaging TOF-SIMS of bone mineralization on porous titanium implants after 1 week in bone, (2006), pp. 6757–6760
26. H. Nygren, C. Eriksson, K. Hederstierna, P. Malmberg, TOF-SIMS analysis of the interface between bone and titanium implants-Effect of porosity and magnesium coating, (2008), pp. 1092–1095
27. E. Acquadro, C. Cabella, S. Ghiani, L. Miragoli, E.M. Bucci, D. Corpillo, Matrix-assisted laser desorption ionization imaging mass spectrometry detection of a magnetic resonance imaging contrast agent in mouse liver. Anal. Chem. **81**, 2779–2784 (2009)
28. S.J. Atkinson, P.M. Loadman, C. Sutton, L.H. Patterson, M.R. Clench, Examination of the distribution of the bioreductive drug AQ4N and its active metabolite AQ4 in solid tumours by imaging matrix-assisted laser desorption/ionisation mass spectrometry. Rapid Commun. Mass Spectrom. **21**, 1271–1276 (2007)
29. L. Signor, E. Varesio, R.F. Staack, V. Starke, W.F. Richter, G. Hopfgartner, Analysis of erlotinib and its metabolites in rat tissue sections by MALDI quadrupole time-of-flight mass spectrometry. J. Mass Spectrom. **42**, 900–909 (2007)
30. P.J. Trim, C.M. Henson, J.L. Avery, A. McEwen, M.F. Snel, E. Claude et al., Matrix-assisted laser desorption/ionization-ion mobility separation-mass spectrometry imaging of vinblastine in whole body tissue sections. Anal. Chem. **80**, 8628–8634 (2008)
31. J.M. Wiseman, D. R. Ifa, Y. X. Zhu, C. B. Kissinger, N. E. Manicke, P.T. Kissinger et al., *Desorption electrospray ionization mass spectrometry: Imaging drugs and metabolites in tissues, Proceedings of the National Academy of Sciences of the United States of America*, vol. 105 (2008), pp. 18120–18125
32. M. Zoriy, A. Matusch, T. Spruss, J.S. Becker, Laser ablation inductively coupled plasma mass spectrometry for imaging of copper, zinc, and platinum in thin sections of a kidney from a mouse treated with cis-platin. Int. J. Mass Spectrom. **260**, 102–106 (2007)
33. T.K. Sinha, S. Khatib-Shahidi, T.E. Yankeelov, K. Mapara, M. Ehtesham, D.S. Cornett et al., Integrating spatially resolved three-dimensional MALDI IMS with in vivo magnetic resonance imaging. Nat. Methods **5**, 57–59 (2008)

34. K.A. Veselkov, R. Mirnezami, N. Strittmatter, R.D. Goldin, J. Kinross, A.V. Speller, et al., Chemo-informatic strategy for imaging mass spectrometry-based hyperspectral profiling of lipid signatures in colorectal cancer. Proc. Natl. Acad. Sci. USA **111**, 1216–21 (2014)
35. G.B. Eijkel, B.K. Kaletas, I.M. van der Wiel, J.M. Kros, T.M. Luider, R.M.A. Heeren, Correlating MALDI and SIMS imaging mass spectrometric datasets of biological tissue surfaces. Surf. Interface Anal. **41**, 675–685 (2009)
36. A. Broersen, R. van Liere, A.F.M. Altelaar, R.M.A. Heeren, L.A. McDonnell, Automated, feature-based image alignment for high-resolution imaging mass spectrometry of large biological samples. J. Am. Soc. Mass Spectrom. **19**, 823–832 (2008)
37. A.C. Crecelius, D.S. Cornett, R.M. Caprioli, B. Williams, B.M. Dawant, B. Bodenheimer, Three-dimensional visualization of protein expression in mouse brain structures using imaging mass spectrometry. J. Am. Soc. Mass Spectrom. **16**, 1093–1099 (2005)
38. L. Shi, G. Campbell, W.D. Jones, F. Campagne, Z. Wen, S.J. Walker et al., The MicroArray Quality Control (MAQC)-II study of common practices for the development and validation of microarray-based predictive models. Nat. Biotechnol. **28**, 827–838 (2010)
39. L. Martens, M. Chambers, M. Sturm, D. Kessner, F. Levander, J. Shofstahl, et al., mzML-a Community Standard for Mass Spectrometry Data. Mol. Cell. Proteomics. **10** (2011)
40. K. Schwamborn, R.C. Krieg, M. Reska, G. Jakse, R. Knuechel, A. Wellmann, Identifying prostate carcinoma by MALDI-imaging. Int. J. Mol. Med. **20**, 155–159 (2007)
41. M. Hanselmann, U. Kothe, M. Kirchner, B.Y. Renard, E.R. Amstalden, K. Glunde et al., Toward digital staining using imaging mass spectrometry and random forests. J. Proteome Res. **8**, 3558–3567 (2009)
42. D.D. Lee, H.S. Seung, Algorithms for non-negative matrix factorization. Adv. Neural. Inf. Process. Syst. **13**, 556–562 (2002)
43. P.W. Siy, R.A. Moffitt, R.M. Parry, Y.F. Chen, Y. Liu, M.C. Sullards et al., *Matrix Factorization Techniques for Analysis of Imaging Mass Spectrometry Data*, (2008)
44. X.C. Xiong, X. Fang, Z. Ouyang, Y. Jiang, Z.J. Huang, Y.K. Zhang, Feature extraction approach for mass spectrometry imaging data using non-negative matrix factorization. Chin. J. Anal. Chem. **40**, 663–669 (2012)
45. E.A. Jones, R. Shyti, R.J.M. van Zeijl, S.H. van Heiningen, M.D. Ferrari, A.M. Deelder et al., Imaging mass spectrometry to visualize biomolecule distributions in mouse brain tissue following hemispheric cortical spreading depression. Journal of Proteomics **75**, 5027–5035 (2012)
46. E.A. Jones, A. van Remoortere, R.J.M. van Zeijl, P.C.W. Hogendoorn, J. Bovee, A.M. Deelder et al., Multiple statistical analysis techniques corroborate intratumor heterogeneity in imaging mass spectrometry datasets of myxofibrosarcoma. Plos One **6** (2011)
47. R.M. Parry, A.S. Galhena, C.M. Gamage, R.V. Bennett, M.D. Wang, F.M. Fernandez, omniSpect: an open MATLAB-based tool for visualization and analysis of matrix-assisted laser desorption/ionization and desorption electrospray ionization mass spectrometry images. J. Am. Soc. Mass Spectrom. **24**, 646–649 (2013)
48. A. Hyvärinen, E. Oja, Independent component analysis: algorithms and analysis. Neural Networks **13**, 411–430 (2000)
49. C.J.C. Burges, Dimension reduction: a guided tour. Found. Trends® Mach. Learn. **2**, 275–365 (2010)
50. L.A. Klerk, A. Broersen, I.W. Fletcher, R. van Liere, R.M.A. Heeren, Extended data analysis strategies for high resolution imaging MS: new methods to deal with extremely large image hyperspectral datasets. Int. J. Mass Spectrom. **260**, 222–236 (2007)
51. S.O. Deininger, M.P. Ebert, A. Futterer, M. Gerhard, C. Rocken, MALDI imaging combined with hierarchical clustering as a new tool for the interpretation of complex human cancers. J. Proteome Res. **7**, 5230–5236 (2008)
52. A.M. Oros-Peusquens, A. Matusch, J.S. Becker, N.J. Shah, Automatic segmentation of tissue sections using the multielement information provided by LA-ICP-MS imaging and k-means cluster analysis. Int. J. Mass Spectrom. **307**, 245–252 (2011)

53. T. Alexandrov, M. Becker, S.O. Deininger, G. Ernst, L. Wehder, M. Grasmair et al., Spatial segmentation of imaging mass spectrometry data with edge-preserving image denoising and clustering. J. Proteome Res. **9**, 6535–6546 (2010)
54. T. Alexandrov, MALDI imaging mass spectrometry: statistical data analysis and current computational challenges. BMC Bioinf. **13**, S11 (2012)
55. P. Berkhin, Survey of clustering data mining techniques. *Accrue Software, Inc. Technical Report,* (2002)
56. L.A. McDonnell, A. van Remoortere, R.J.M. van Zeijl, A.M. Deelder, Mass spectrometry image correlation: quantifying colocalization. J. Proteome Res. **7**, 3619–3627 (2008)
57. C.D. Kaddi, R.M. Parry, M.D. Wang, Hypergeometric similarity measure for spatial analysis in tissue imaging mass spectrometry. *Proceedings of IEEE BIBM 2011*, pp. 604–607, (2012)
58. C.D. Kaddi, R.M. Parry, M.D. Wang, Multivariate hypergeometric similarity measure. IEEE/ACM Trans. Comput. Biol. Bioinform. **10**, 1505–16 (2013)
59. R. van de Plas, K. Pelckmans, B. De Moor, W. E., Spatial querying of imaging mass spectrometry data: a nonnegative least squares approach. Neural Inf. Proc. Syst. Workshop Mach. Learn. Comput. Biol. (2007)

Identification and Functional Annotation of LncRNAs in Human Disease

Qi Liao, Dechao Bu, Liang Sun, Haitao Luo and Yi Zhao

Abstract Accumulated evidence suggests that long noncoding RNAs (lncRNAS) play a key role in most of the biological processes. By the advance of sequencing technology, more and more lncRNAs are identified. However, only a few has known functions. It is still a challenge to annotate the functions of lncRNAs for both bioinformaticians and biologists. In this chapter, we gave a comprehensive review of the current bioinformatics methods to identify lncRNAs and annotate their functions in mammals. The identification of lncRNAs was mainly based on the technologies of microarray and RNA-seq. While for the functional annotations of lncRNAs, a method based on the co-expression network of both coding and non-coding genes was illustrated. We also reviewed several ways to analyze the interactions between lncRNAs and targets such as miRNAs and protein-coding genes. An example of identifying and annotating human lncRNAs was given to illustrate the whole process.

Background

Long noncoding RNAs (lncRNAs) are a kind of noncoding RNAs with lengths longer than 200nt [1]. LncRNAs were regarded as mRNA-like ncRNAs for a period as their sequence characters are similar to mRNAs, e.g., being transcribed from RNA poly II, splicing, acquiring poly-A tails and 5'caps except for encoding proteins, etc. [1]. During the last several years, numerous lncRNAs were identified in mammalian genomes through cDNA sequencing, RNA-seq and computational methods [2–6]. These lncRNAs, together with protein-coding genes, constitute the complex architecture of

Q. Liao (✉) · D. Bu · L. Sun · H. Luo · Y. Zhao
Key Lab of Intelligent Information Processing of Chinese Academy of Sciences (CAS), Institute of Computing Technology, CAS, 100190 Beijing, China
e-mail: liaoqi@nbu.edu.cn

Q. Liao
Department of Preventative Medicine, School of Medicine, Ningbo University, 315211 Ningbo, Zhejiang, China

D. Xu et al. (eds.), *Health Informatics Data Analysis*, Health Information Science,
DOI 10.1007/978-3-319-44981-4_3

mammalian genome [3]. For example, the lncRNAs located in intronic, intergenic regions or the exons of protein-coding gene regions with converse transcription direction are called intronic lncRNAs, large intergenic long noncoding RNAs (lincRNAs and antisense lncRNAs, respectively). A considerable number of studies reported that these lncRNAs play an important role in the regulation of protein-coding genes including modifying the expression, activity and location of protein-coding genes [7]. Therefore, lncRNAs participate in a variety of biological processes such as development, gene imprinting, immune response, and so on [7–9].

The essential role of lncRNAs in the molecular biological processes attracts researchers to investigate whether the lncRNAs have differential expressions in human diseases, compared with the healthy controls. Some of them may serve as biomarkers for disease diagnosis and prognosis. For example, upregulated expression of the lncRNA MALAT1 was considered as a biomarker of poor prognosis in colorectal cancer [10], while decreased expression of the lncRNA GAS5 also indicated a poor prognosis in gastric cancer [11]. Although quite a few novel lncRNAs have been identified in various human diseases, the biological functions of most lncRNAs remain unknown. Through decades of efforts, scientists have developed some efficient technologies to detect disease-associated lncRNAs. This chapter will introduce the state-of-the-art technologies to identify novel lncRNAs and annotated their functions in human diseases.

Current Bioinformatics Methods

Identify Associated lncRNAs in Human Diseases

By Microarray

Microarray is a widely used method to simultaneously detect the expression levels of all genes (protein-coding genes in usual) in a whole genome. Based on the technology of microarray, researchers are able to find differently expressed protein-coding genes in a variety of biological processes associated with human diseases. Microarrays can be classified as DNA microarray, RNA microarray, and protein microarray, etc. The major microarray service providers include Affymetrix, Agilent, and Illumina. The RNA microarray from the Affymetrix company was designed based on the sequence databases of Refseq, EST, and Unigene. We have observed that some probes corresponding to EST sequences in the Affymetrix microarrays match perfectly with known lncRNAs [12]. Therefore, by re-annotating the probes of the Affymetrix microarrays, the expression signals of these probe-matching lncRNAs may have already been screened in the publicly available microarray datasets.

In order to get the expression profiles of lncRNAs, first we need to create a new Chip Description File (CDF) in which probes of both protein-coding genes and lncRNAs are included. That is to say, the probe sequences are annotated to match lncRNAs as well as protein-coding genes using alignment tools like BLASTn. The alignment results are filtered by the following steps: (1) Only probes perfectly matched to known transcripts will be retained. (2) All transcripts corresponding to the retained probes should be mapped to the genome and annotated on the gene level. For example, the Refseq transcripts should be mapped to the Entrez Gene. (3) Genes matching fewer than three probes are excluded. By the above three steps, a new CDF package is created in R software, which is a statistical analysis program and is frequently used in the bioinformatics data analysis.

So the new CDF annotation file translates the expression profile of only protein-coding genes in the GEO database into an expression profile of both lncRNAs and protein-coding genes. By analyzing the microarray datasets of human disease samples, we can detect the expression levels these annotated lncRNAs and investigate the differentially expressed lncRNAs associated with human diseases. We have created a web server called ncFANs to calculate the coding and noncoding gene expression profiles and to identify aberrantly expressed both protein-coding

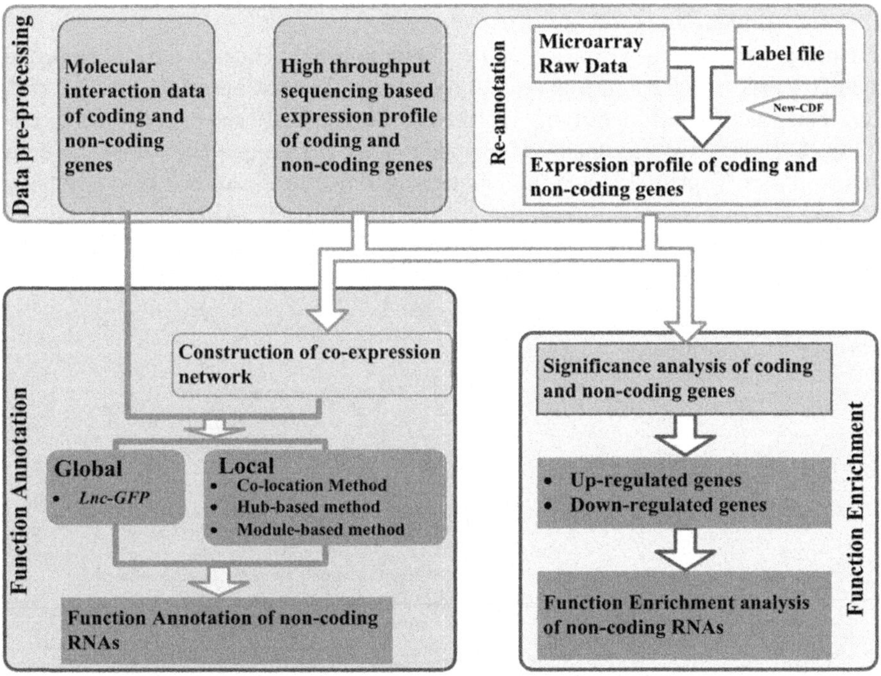

Fig. 1 The workflow of ncFANs. The ncFANs server consists of two parts: data preprocessing and lncRNA function annotation. Data preprocessing mainly transforms the original microarray expression profiles into the coding noncoding gene expression profiles. While function annotation provides two strategies: based on co-expression network and finding aberrantly expressed genes including both lncRNAs and protein-coding genes. By use of ncFANs, users can select a microarray dataset of Affymetrix in certain disease from GEO database, and find the differently expressed lncRNAs or predict the functions of associated lncRNAs in human disease. The figure is derived from the ncFANs website (http://www.bioinfo.org/ncfans/about.php)

genes and lncRNAs (Fig. 1, http://www.bioinfo.org/ncfans/) [18]. For example, GSE19826 is an expression dataset with different stages of gastric cancer and corresponding adjacent tissue [13]. If we format it as the expression profiles of protein-coding genes and lncRNAs using the ncFANs server, we can apply *t*-test or fold change analysis to find the differentially expressed lncRNAs in different development stages of gastric cancer. A set of examples for uploading files and preprocessing tools are also provided in ncFANs (http://www.bioinfo.org/ncfans/download.php).

By RNA-seq

RNA-seq is a recently developed technology for deep profile the transcriptome including mRNA transcripts and other different populations of RNAs such as miRNAs and tRNAs. So RNA-seq can detect the expression levels of both protein-coding genes and lncRNAs. Many novel lncRNAs have been identified in mammals such as humans and other species at an accelerated speed using the RNA-seq technology [4–6, 14, 15].

During the last few years, several computational pipelines were developed to identify novel lncRNAs in biological processes of interest based on the RNA-seq data. The procedure of identifying lncRNAs using the RNA-seq data is summarized in Fig. 2. First, RNA-seq reads are mapped to the reference genome by the computer program Tophat [16] and the uniquely matched reads are de novo assembled into transcripts by the computer program Cufflinks [17]. After filtering out the transcripts with low coverage, the remaining transcripts are merged using the

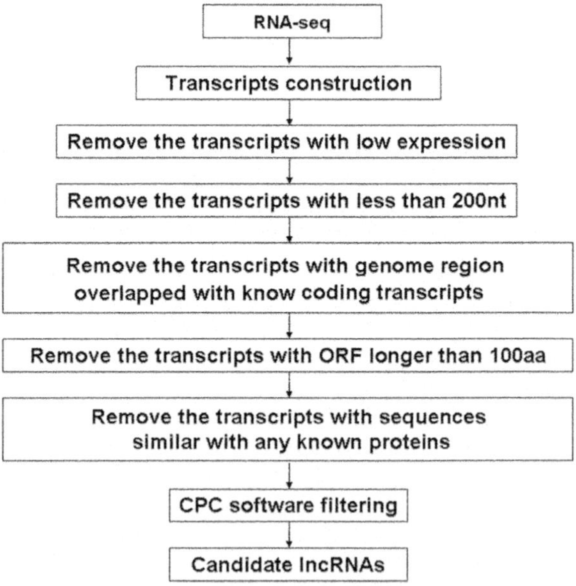

Fig. 2 The pipeline of identifying novel lncRNAs from RNA-seq dataset

computer program Cuffmerge. Then the transcripts are selected as candidate lncRNAs if they do not overlap with the genomic regions of known protein-coding genes and other noncoding RNAs.

Finally, the following four steps are employed to determine whether a transcript is a lnRNA. (1) The length of the transcript is longer than 200 nt. (2) The length of the predicted ORFs by ORFfinder or other ORF prediction programs is not longer than 100 aa on either the positive or the negative strand. (3) The transcript sequence is not homologous to any known proteins, using the BLAST program and 30% as the cutoff for both the similarity and identity length ratios. (4) The transcript is predicted to be "noncoding" using the Codon Potential Score (CPC) program (available in web server, http://cpc.cbi.pku.edu.cn/) with default parameters. The coding potentiality may also be predicted using other similar programs. The lncRNA annotations in human disease RNA-seq datasets demonstrate accurate prediction and functional characterization of lncRNAs associated with human diseases.

Annotated the Functions of Associated lncRNAs in Human Diseases Based on Co-expression Network

If multiple time points or experimental conditions are available, we may annotate the functions of novel lncRNAs associated with human diseases through co-expression network analysis. First, we may construct a gene co-expression network. The Pearson correlation coefficient (Pcc) is frequently employed to measure how a pair of genes is co-expressed. The Pcc p-value needs to be adjusted by the Bonferroni or other statistical methods. Then the co-expression relationships with adjusted P-values of smaller than a cutoff (0.05 or 0.01) are selected to construct a gene co-expression network. If several datasets are available, the overall co-expression relationship may be determined by whether the co-expression relationship between two genes is detected in at least a certain number of dataset.

When the co-expression network is constructed, genomic co-location, hub-based, and module-based methods can be utilized to predict the functions of novel lncRNAs within the network. These three methods have been integrated with the co-expression network to annotate the functions of novel lncRNAs in the mouse genome and were comprehensively verified by different technologies [12]. The genomic co-location refers to two genes that are co-expressed and spaced by at most 100 kb in their genomic locations. If a lncRNA is co-expressed and co-located with a nearby protein-coding gene, the lncRNA may regulate the protein-coding gene which has similar functions. In the hub-based method, lncRNAs are predicted based on the functions of first-level associated protein-coding genes with known Gene Ontology (GO) annotation, including GO Biological Process (BP), Cellular Component (CC), and Molecular Function (MF). The enriched GO terms calculated in the set of near coding genes were annotated to the lncRNA in the center node. In the module-based method, the MCL algorithm can be used to mine out the modules of the co-expression network. Then, for each module, the method similar to hub-based method can be used to obtain enriched functions to annotate the functions of lncRNAs within the module. In order to conveniently annotate the functions of lncRNAs for biological researchers, we have developed a web server called ncFANs [18].

Further Analysis of LncRNAs After Identification and Functional Annotation

LncRNAs may interact with other large molecules such as protein-coding genes and miRNAs through various mechanisms. After the identification and annotation of lncRNAs, we may want to know the targets of lncRNAs or how lncRNAs are regulated. There are several scientific questions to be investigated, after the detection of novel lncRNAs.

(1) Identification of transcription factor (TF) binding sites of lncRNAs.
The transcription factor (TF) is a protein that binds to the promoter of a protein-coding gene and regulates the expression levels of this gene. LncRNAs may be regulated by TFs in the same way. By screening for the binding sites of some TFs using the biological technologies like ChIP-chip or ChIP-seq, we may also detect the transcriptional regulation relationships between TFs and lncRNAs in the genome scale. For example, Yang JH et al. identified tens of thousands of TF-lncRNA regulatory relationships and developed a database called ChIPBase to visualize the data [19].

(2) Identification of miRNA-lncRNA interaction.
MiRNAs regulate the expression levels of protein-coding genes by targeting the 3' Un-Translated Regions (UTRs) of the genes' mRNAs. The sequences of miRNA and the targeting UTR may be partially matched in mammals or perfectly identical in plants. Genome-wide screening of argonaute proteins binding genomic regions found that 5% were located in the noncoding genes [20], so some lncRNAs may be regulated by miRNAs. The regulatory relationship between microRNAs and lncRNAs may be predicted using the miRNA target prediction programs or co-expression relationships. For example, Liran Juan et al. found 90 pairs of lncRNA-miRNAs regulations with strong reverse expression correlation [21].

(3) Target identification of lncRNAs acting as ceRNAs.
Competing endogenous RNAs (ceRNAs) regulate the targets by competing for the binding sites of miRNAs. MiRNAs can regulate both protein-coding genes and lncRNAs, and lncRNAs may act as ceRNAs to regulate the targets who share the same microRNA target sites. For example, the lncRNA HOTAIR regulates the expression of HER2 in gastric cancer by acting as ceRNA [22]. And the lncRNA linc-MD1 can act as a ceRNA to regulate MAML1 and MEF2C to control the differentiation of muscle [23]. Based on the relationships

between microRNA-lncRNAs and microRNA-mRNAs, we can predict lncRNA(ceRNA)-target interactions. For example, Li et al. predicted ~10,000 ceRNA-target interactions based on the 108 CLIP-seq datasets and developed a web server called ceRNAFunction (http://starbase.sysu.edu.cn/ceRNAFunctio.php) [24]. Das et al. constructed a database called lncCeDB to record the lncRNAs acting as ceRNAs (http://gyanxet-beta.com/lncedb/) [25].

Challenges and Current Problems

1. LncRNAs are abundant in mammalian genomes, and usually do not encode peptides. The current technologies do not work well on determining the exact transcript start site, transcript terminal site, and splice site of the lncRNAs.
2. Microarray-based revised annotation strategy can only identify a limit number of lncRNAs that match to the designed probes in the microarray platform.
3. Although a number of computational pipelines were developed to identify lncRNAs from the assembled transcripts of RNA-seq data, the prediction accuracy remains to be improved. More accurate and efficient program is needed to detect novel lncRNAs by integrating the OMIC data.
4. Accurate functional annotations of lncRNAs are also in urgent need to develop.

Example

Identification of Differentially Expressed lncRNAs in Gastric Cancer

(1) **Data**
The gastric cancer microarray dataset was profiled using the Affymetrix platform HgU133Plus2, and downloaded from the database GEO with the ID GSE19826. The dataset consists of four different stages (I, II, III and IV) of gastric cancer samples and the paired adjacent noncancer tissues. Each sample has three replicates.

(2) **Method**
The raw data was downloaded from the database GEO (http://www.ncbi.nlm.nih.gov/geo/query/acc.cgi?acc=GSE19826) and preprocessed in a local computer using the R script downloaded from the web server ncFANs. Then the pre-processed profile was uploaded to the web server ncFANs to analyze the gene expression profile and detect differentially expressed lncRNAs between different cancer stages. *T*-test was utilized to measure the differential significance, and the cutoff of *P*-value was set as 0.05.

(3) **Results**
The numbers of differentially expressed protein-coding genes and lncRNAs are shown in Tables 1 and 2. We found that MALAT1 was upregulated in the cancer stage III compared against the paired adjacent tissues. It has been reported that MALAT1 was abnormally highly expressed in gastric cancer cell lines previously [26]. These differentially expressed lncRNAs may represent complementary diagnosis biomarkers in gastric cancer.

Table 1 The numbers of differentially expressed protein-coding genes and lncRNAs between different stages of gastric cancer

	Downregulated lncRNAs	Downregulated code genes	Upregulated lncRNAs	Upregulated code genes
Stage I versus Stage II	12	154	11	219
Stage I versus Stage III	19	111	6	69
Stage I versus Stage IV	24	172	7	88
Stage II versus Stage III	45	251	5	134
Stage II versus Stage IV	35	219	6	92
Stage III versus Stage IV	7	106	16	91

Table 2 The numbers of differentially expressed protein-coding genes and lncRNAs comparing with paired adjacent tissues in different stages of gastric cancer

	Downregulated lncRNAs	Downregulated code genes	Upregulated lncRNAs	Upregulated code genes
Stage I	7	204	37	370
Stage II	13	108	18	146
Stage III	8	95	18	105
Stage IV	15	292	22	366

Conclusion

LncRNAs regulate the protein-coding genes through diverse mechanisms. Some known lncRNAs were observed to be statistically significantly associated with human diseases, but many more lncRNAs remain to be functionally characterized.

Re-annotated microarray profile and RNA-seq based detection are two major bioinformatics strategies to identify novel lncRNAs. Researchers are actively working on more efficient and accurate technologies to screen the novel lncRNAs in the mammalian genomes.

References

1. V.A. Erdmann, M. Szymanski, A. Hochberg, N. Groot, J. Barciszewski, Non-coding, mRNA-like RNAs database Y2K. Nucleic Acids Res. **28**, 197–200 (2000)
2. T. Ota, Y. Suzuki, T. Nishikawa, T. Otsuki, T. Sugiyama et al., Complete sequencing and characterization of 21,243 full-length human cDNAs. Nat. Genet. **36**, 40–45 (2004)
3. Y. Okazaki, M. Furuno, T. Kasukawa, J. Adachi, H. Bono et al., Analysis of the mouse transcriptome based on functional annotation of 60,770 full-length cDNAs. Nature **420**, 563–573 (2002)
4. M.A. Hassan, M.B. Melo, B. Haas, K.D. Jensen, J.P. Saeij, De novo reconstruction of the Toxoplasma gondii transcriptome improves on the current genome annotation and reveals alternatively spliced transcripts and putative long non-coding RNAs. BMC Genom. **13**, 696 (2012)
5. M. Guttman, M. Garber, J.Z. Levin, J. Donaghey, J. Robinson et al., Ab initio reconstruction of cell type-specific transcriptomes in mouse reveals the conserved multi-exonic structure of lincRNAs. Nat. Biotechnol. **28**, 503–510 (2010)
6. J.W. Nam, D.P. Bartel, Long noncoding RNAs in C. elegans. Genome Res. **22**, 2529–2540 (2012)
7. J.E. Wilusz, H. Sunwoo, D.L. Spector, Long noncoding RNAs: functional surprises from the RNA world. Genes Dev. **23**, 1494–1504 (2009)
8. R.J. Taft, K.C. Pang, T.R. Mercer, M. Dinger, J.S. Mattick, Non-coding RNAs: regulators of disease. J. Pathol. **220**, 126–139 (2010)
9. T.R. Mercer, M.E. Dinger, J.S. Mattick, Long non-coding RNAs: insights into functions. Nat. Rev. Genet. **10**, 155–159 (2009)
10. H.T. Zheng, D.B. Shi, Y.W. Wang, X.X. Li, Y. Xu et al., High expression of lncRNA MALAT1 suggests a biomarker of poor prognosis in colorectal cancer. Int. J. Clin. Exp. Pathol. **7**, 3174–3181 (2014)
11. M. Sun, F.Y. Jin, R. Xia, R. Kong, J.H. Li et al., Decreased expression of long noncoding RNA GAS5 indicates a poor prognosis and promotes cell proliferation in gastric cancer. BMC Cancer **14**, 319 (2014)
12. Q. Liao, C. Liu, X. Yuan, S. Kang, R. Miao et al., Large-scale prediction of long non-coding RNA functions in a coding-non-coding gene co-expression network. Nucleic Acids Res. **39**, 3864–3878 (2011)
13. Q. Wang, Y.G. Wen, D.P. Li, J. Xia, C.Z. Zhou et al., Upregulated INHBA expression is associated with poor survival in gastric cancer. Med. Oncol. **29**, 77–83 (2012)
14. L. Sun, Z. Zhang, T.L. Bailey, A.C. Perkins, M.R. Tallack et al., Prediction of novel long non-coding RNAs based on RNA-Seq data of mouse Klf1 knockout study. BMC Bioinform. **13**, 331 (2012)
15. S. Ren, Z. Peng, J.H. Mao, Y. Yu, C. Yin et al., RNA-seq analysis of prostate cancer in the Chinese population identifies recurrent gene fusions, cancer-associated long noncoding RNAs and aberrant alternative splicings. Cell Res. **22**, 806–821 (2012)
16. C. Trapnell, L. Pachter, S.L. Salzberg, TopHat: discovering splice junctions with RNA-Seq. Bioinformatics **25**, 1105–1111 (2009)

17. C. Trapnell, B.A. Williams, G. Pertea, A. Mortazavi, G. Kwan et al., Transcript assembly and quantification by RNA-Seq reveals unannotated transcripts and isoform switching during cell differentiation. Nat. Biotechnol. **28**, 511–515 (2010)
18. Q. Liao, H. Xiao, D. Bu, C. Xie, R. Miao et al., ncFANs: a web server for functional annotation of long non-coding RNAs. Nucleic Acids Res. **39**, W118–W124 (2011)
19. J.H. Yang, J.H. Li, S. Jiang, H. Zhou, L.H. Qu, ChIPBase: a database for decoding the transcriptional regulation of long non-coding RNA and microRNA genes from ChIP-Seq data. Nucleic Acids Res. **41**, D177–D187 (2013)
20. S.W. Chi, J.B. Zang, A. Mele, R.B. Darnell, Argonaute HITS-CLIP decodes microRNA-mRNA interaction maps. Nature **460**, 479–486 (2009)
21. L. Juan, G. Wang, M. Radovich, B.P. Schneider, S.E. Clare et al., Potential roles of microRNAs in regulating long intergenic noncoding RNAs. BMC Med. Genomics **6**(Suppl 1), S7 (2013)
22. X.H. Liu, M. Sun, F.Q. Nie, Y.B. Ge, E.B. Zhang et al., Lnc RNA HOTAIR functions as a competing endogenous RNA to regulate HER2 expression by sponging miR-331-3p in gastric cancer. Mol. Cancer **13**, 92 (2014)
23. M. Cesana, D. Cacchiarelli, I. Legnini, T. Santini, O. Sthandier et al., A long noncoding RNA controls muscle differentiation by functioning as a competing endogenous RNA. Cell **147**, 358–369 (2011)
24. J.H. Li, S. Liu, H. Zhou, L.H. Qu, J.H. Yang, starBase v2.0: decoding miRNA-ceRNA, miRNA-ncRNA and protein-RNA interaction networks from large-scale CLIP-Seq data. Nucleic Acids Res. **42**, D92–D97 (2014)
25. S. Das, S. Ghosal, R. Sen, J. Chakrabarti, lnCeDB: database of human long noncoding RNA acting as competing endogenous RNA. PLoS ONE **9**, e98965 (2014)
26. J. Wang, L. Su, X. Chen, P. Li, Q. Cai et al., MALAT1 promotes cell proliferation in gastric cancer by recruiting SF2/ASF. Biomed. Pharmacother. **68**, 557–564 (2014)

Metabolomics Characterization of Human Diseases

Masahiro Sugimoto

Abstract Recent *omics* technologies have realized the comprehensive identification and quantification of metabolites, named metabolomics. Mass spectrometry with molecular separation system is commonly used and hundreds of metabolite concentrations should be dealt with in this field. To help interpreting these data, the development of variety of data processing tools, database resources, visualization software, and pathway analyses are still active. Here, we review the typical data analyses of metabolomics data and introduce step-by-step tutorials of (1) metabolic pathway analysis to understand aberrance of multiple metabolites in individual pathways, and (2) development and validation of a discrimination model using multiple markers to differentiate patients with diseases from healthy controls. These tools and protocols are versatile and can be used for the analyses for any other diseases and datasets.

Background

Omics technologies provide a holistic view of the complex interactions between thousands of molecules within a biological system. Metabolomics, one of the more recent *omics* technologies, deals with the quantitative global profiling of metabolites. Although analytical and bioinformatics technologies in genomics, transcriptomics, and proteomics are well established and widely used in biological and medical researches, these approaches do not provide a complete picture. Metabolomics is considered to fill a gap between phenotype and genotype, more directly reflecting the immediate status of a biological system because metabolite profiles are controlled by all of the upstream information contained within the "central dogma". Thus, metabolomics has been used to explore the dynamic

---

M. Sugimoto (✉)
Institute for Advanced Biosciences, Keio University, 246-2 Mizukami, Kakuganji, Tsuruoka, Yamagata 997-0052, Japan
e-mail: msugi@sfc.keio.ac.jp

D. Xu et al. (eds.), *Health Informatics Data Analysis*, Health Information Science,
DOI 10.1007/978-3-319-44981-4_4

response of living systems under diverse physiological and pathological conditions, and to characterize human disease.

Among multiple analytical platforms for profiling metabolites, nuclear magnetic resonance (NMR) and mass spectrometry (MS) are the major types currently available. One clear advantage of NMR is its ability to measure biological samples in a nondestructive manner; it can thus provide a wealth of biochemical information not observable by MS. NMR is therefore currently the most popular method being used in half of the reported metabolomics studies [1]. Separation systems such as gas chromatography (GC), liquid chromatography (LC), and capillary electrophoresis (CE) are usually used prior to MS for separating molecules by chemical features. These separation–detection approaches, known as hyphenated MS technologies, are becoming dominant because of their higher sensitivity and separation abilities compared with NMR. However, because of the large diversity of chemical features in metabolites, no single analytical approach can profile all metabolites comprehensively [2]. GC–MS is the most well established among the hyphenated MS systems and can identify and quantify hundreds of volatile metabolites. LC–MS has rapidly become prevalent in metabolomics for profiling a wider range of metabolites, such as sugars, lipids, amino acids, and a variety of secondary metabolites usually observed in plants. CE–MS is still minor but has an ability to separate and quantify hundreds of charged metabolites simultaneously using only two charge (positive and negative) modes. Data processing pipelines starting from file conversion of raw data to generating a metabolite concentration matrix (metabolite $\times$ samples) depend on each individual approach, or on the technology. The individual techniques and variety of software tools for data processing have been reviewed elsewhere [3–5]. In this chapter, subsequent bioinformatics analyses that are commonly used, and can be applied independently to each measurement approach, are described. Of particular relevance to characterizing human diseases based on their metabolomics profiles, two typical data analysis flows and step-by-step procedures to use several software tools are introduced.

Challenges

To use metabolomics for understanding human diseases, two typical scenarios are considered. The first is to identify a metabolic aberrance (usually in data obtained from cultured cells or organs) based on information regarding metabolic pathways, i.e., pathway analysis. The second is to identify the ability of metabolites to discriminate depending on the phenotypes of the measured samples, i.e., biomarker discovery. The development of classification model using multiple metabolite markers is included in this scenario. In addition to these analyses, interpretation of interesting metabolites is performed using several databases that include relevant information.

(a) **Pathway analysis**

When comparing metabolomics profiles between two phenotypes, e.g., with and without diseases, there are rare cases in which a single metabolite shows a major difference between the two groups. When this is not the case, a consistent change of several metabolites occurring in a functionally categorized pathway, such as glycolysis, the citric acid cycle, or the pentose phosphate pathway, can be helpful to understand changes of metabolic profile patterns. The interactions of many metabolites with their regulatory elements, such as enzymes and transporters, have already been well documented and are available in several public and commercial databases. In a typical pathway analysis, quantified metabolite concentrations are projected onto these metabolic pathways to identify the aberrant metabolic pathways. Such an approach is now frequently used for the analysis of metabolic data in cultured cell or tissue samples [6].

(b) **Biomarker discovery**

Metabolomics profiles of biofluids obtained in a minimally invasive manner, such as blood, urine, and saliva, have the potential to differentiate specific diseases. Thus, biomarker discovery, or the development of classification models, has frequently been used to diagnose diseases. For example, GC–MS profiles of serum metabolites have been used to diagnose pancreatic cancer cases from healthy controls [7]. A combination of multiple salivary metabolites differentiated between oral cancers and healthy controls [8]. These applications were analyzed using multivariate analyses such as principal component analysis (PCA) and multiple logistic regression (MLR).

Current Techniques

Analysis techniques commonly used for metabolomics data are introduced here.

(a) **Data visualization to understand metabolic profiles**

To visualize overall data, PCA is probably the most widely used in metabolomics studies. This converts high-dimensional data into fewer dimensions, maintaining as much variance as possible from the original data. The data plots distributed on the principal component (PC) space help to understand the similarity of metabolite patterns among samples, especially for samples representing multiple phenotypes (≥ 3). This is not a classification method *per se*, but PC values are frequently used to differentiate samples with/without diseases. For example, a PCA model consisting of GC/MS metabolite profiles differentiated gastric cancer patients from healthy controls [9]. This method is used to understand the correlation between sample and phenotype, and also to detect outlier samples that show widely different patterns compared with the majority of samples.

Data visualization in a pathway form facilitates understanding of metabolomic aberrance due to disease. MetaboAnalyst [10, 11], Pathway Projector [12], and KEGG [13] provide web-based data mapping functions on metabolic pathways. KEGG provides wide range of pathways in various species including human as well as various data accessible ways, which facilitates the utility of registered data through the users' programs. Pathway Projector utilizes this function and enables the visualization of the given metabolic data in a large pathway. MetaboAnalyst provides integrated data analytical environment, including data visualization and pathway analysis. Vanted, an application that works on a local machine, has a canvas on which users can edit any metabolic pathway to visualize experimental data [14]. Metabolite set enrichment analysis (MSEA) [15] is becoming popular for identifying aberrance not of single metabolites, but of pathways including multiple metabolites. This idea was inspired by gene set enrichment analysis (GSEA), which identifies the enrichment of gene sets belonging to a specific ontology, a technique widely used in the analysis of transcriptomics data [16].

(b) **Classification methods to differentiate samples based on metabolomics profiles**

Partial least squares (PLS) is a regression-based supervised classification method, while PCA is an unsupervised method. PLS suits the analysis of relatively few samples compared with the number of observable features, and is therefore commonly used for developing classification models based on metabolomics profiles. PLS-discriminant analyses (PLS-DA) have been widely used to discriminate between two phenotypes, e.g., samples with diseases from healthy controls. Random forests (RF), a new machine learning method integrating multiple decision trees, is also frequently used for the same purpose [17], e.g., a RF approach using blood lipid metabolite profiles was used for discriminating pancreatic cancers from healthy controls [18]. Decision trees are also used for the development of classification models [19]. A multivariate statistical analysis tool, MLR models, has also been used frequently. Usually a minimal number of independent metabolites are selected and used for the model's inputs. Therefore, only a small number of metabolites need to be quantified, which is a definitive advantage for clinical use of this approach. For example, serum metabolites profiled by GC/MS were analyzed by MLR to discriminate pancreatic cancer from chronic pancreatitis and healthy controls [20]. MLR models using serum short peptides, such as γ-glutamyl dipeptides, profiled by LC–MS/MS are able to differentiate between a variety of liver diseases, including drug-induced liver injury, asymptomatic hepatitis B virus infection, chronic hepatitis B, cirrhosis type C, and hepatocellular carcinoma [21].

(c) **Interpretation databases for the metabolite of interest**

Once metabolites of interests are found, users will investigate their biological functions reported in the literature and can obtain information on related entities, such as precursor and product metabolites, enzymes, regulatory factors, and transporters. KEGG [13], MetaCyc [22], SMPDB [23], and Reactome [24] include these data. The former two include information on a variety of species while the latter two include human-specific datasets. As an example of a commercial database, MetaCore [25] contains metabolites and their regulatory factors retrieved from a number of published papers.

For metabolomics analysis, detected peaks are usually annotated by matching of standard compounds or registered information in public database. To assign possible candidate metabolites to these peaks, several databases provide a library of mass spectra. HMDB stores a number of raw mass spectrum data and provides mass spectrum data matching functions [26]. Mass Bank also stores a range of mass spectrum data obtained from a variety of MS platforms under different measurement conditions [27]. We have developed the MMMDB to provide both CE–MS spectra and metabolite profiles, i.e., sets of metabolite concentrations obtained from multiple tissues from single mice to understand the balance of metabolite concentrations in metabolic pathways among these tissues [28].

(d) **Simulation analysis in systems biology**

To understand the complicated interactions among metabolites in biological systems, mathematical modeling with simulation is the most advanced sophisticated analytical way. One approach uses only static information, i.e., flux of metabolites under steady-state condition. Usually, isotope-labeled metabolites were injected into cultured cells and flux was estimated based on the transition of labeled metabolite, by utilizing the ability of MS to differentiate labeled and unlabeled metabolites [29, 30]. Meanwhile, the other approach utilized dynamic information, i.e., the kinetics of metabolic reactions, e.g., Michaelis–Menten equations [31]. As an example, metabolic pathway in erythrocytes were mathematically modeled using kinetic parameters collected from various literatures and compared the simulated and the experimentally observed time courses of multiple metabolites in primary pathways [32]. Both analyses require highly reproducible and accurately quantified values of individual metabolites in the pathway of interests with correct assignment of metabolite names; however, such simulations would deeply contribute to the understanding of interactions in the given biological systems.

Example One—Pathway Analysis To Understand the Change of Pattern in Metabolomics Profiles

As a first example, pathway analysis of metabolomics data using the MSEA website (http://www.msea.ca/MSEA/) [15], which can analyze data obtained from human samples, is shown. The step-by-step operation starting from preparing an input file to visualizing results is described in Figs. 1 and 2. The procedure to use this analysis tool using sample data [available at the MSEA website (Fig. 1c)] is described here. This sample file is used for example one, and (with slight

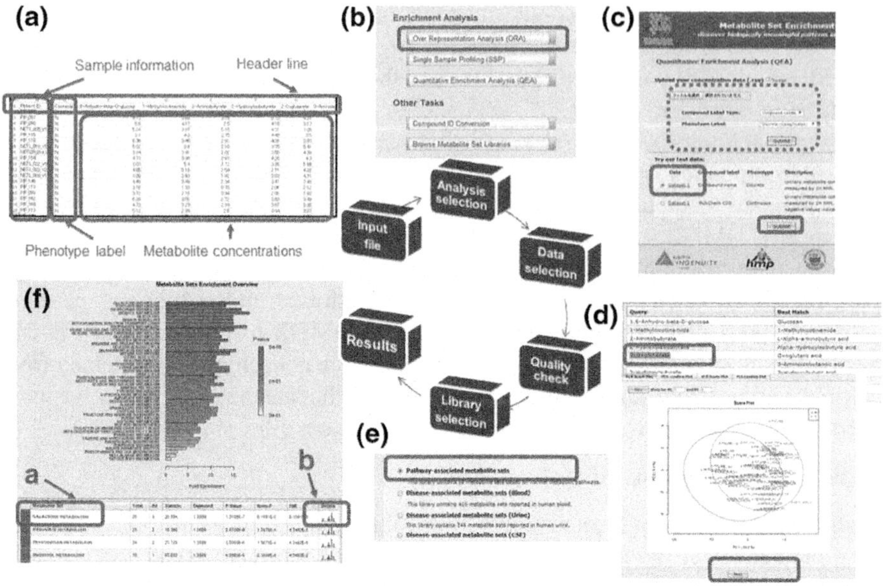

Fig. 1 Analytical flow for MSEA. **a** A data file (csv format) used for MSEA is available as an example at the website (**c**). The first column includes sample information, such as the patient ID. The second column includes a phenotype label. For example, labels "Y" and "N" are used for samples obtained from patients with or without cancer, respectively. The other columns include metabolite concentrations; either absolute concentration or relative quantified values are acceptable here. The first row of each column with metabolite concentration data contains the relevant metabolite name. **b** Selection of analysis type. For this example, the button labeled "Quantitative Enrichment Analysis (QEA)" would be selected. **c** Input file and data type selection. Two data examples are provided here; the first one labeled "Dataset 1" should be selected, followed by the "Submit" button. In the case of uploading the user's own data, the "File select" button shown in the *dashed circle* should be used. **d** Quality check for uploaded data. First, all metabolite names in the upload file are compared with those in the MSEA database, and the metabolites that do not match any entries are highlighted in *red*. Ambiguously matched metabolite names are highlighted in *yellow* and approximate matches are indicated instead. A list of candidate metabolite names is displayed when the highlighted metabolite is clicked. After confirming all such ambiguous metabolites, the "Submit" button at the *bottom* of the website is selected. Overall data are visualized by PCA and PLS. Subsequently, the "Next" button should be selected. **e** Metabolite library selection. The first button, "Pathway-associated metabolite sets", is selected and the "Submit" button at the *bottom* of website clicked. **f** Result overview page. The pathway names are listed according to their enrichment ranking. The *bar graph* indicates both fold enrichment and statistical significance of individual pathways. The *bottom* table has links to show the detail of individual pathways. The metabolite names (*a*) and details icons (*b*) are linked to Fig. 2a, b, respectively

modification) for example two. The data include metabolite concentrations and phenotype labels.

Input data: A data matrix including metabolite concentrations (sample names × metabolite names) must be prepared (Fig. 1a). Two types of phenotype

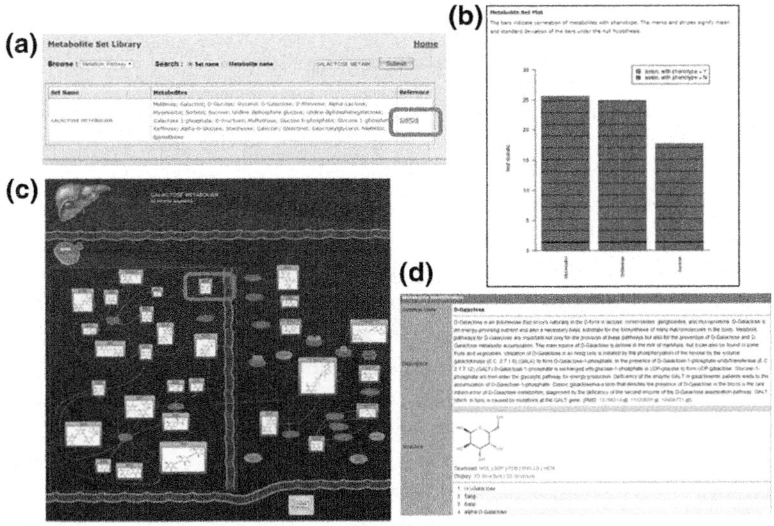

Fig. 2 Results of pathway analysis using MSEA. The panels are fully linked to the output table shown in Fig. 1f, which describes the enriched pathway list. **a** Pathway name and metabolite names. The "SMPDB" at the right column is linked to the pathway database in SMPDB (see *panel c*). **b** Enrichment levels of individual metabolites in a pathway. In this case, three metabolites in the example file (uploaded as described in the legend to Fig. 1) matched metabolites in the galactose pathway (myoinositol, D-glucose, and sucrose). Myoinositol showed the largest fold change difference between the two phenotypes. **c** Pathway visualization in SMPDB. Individual metabolites (*circled* on the figure) are linked to the appropriate entry in the HMDB. **d** Description of an individual metabolite registered in HMDB

for characterizing samples are acceptable; discrete (e.g., healthy or patient in a binary case) or continuous (e.g., body mass index).

What is available after the analysis?: A list of enriched pathways is obtained. These pathways are ranked based on their level of enrichment, identifying which pathways show a strong correlation with the given phenotype. The results website is well linked to relevant information, facilitating the collection of suitable literature.

Overall analytical flow: Users first select the type of analysis (Fig. 1b). A data file is then uploaded and a quality check of the data is conducted (Fig. 1c, d). Subsequently, the metabolite library or libraries to be used for the analysis(such as blood, urine, and cerebral spinal fluid (CSF) metabolites)are selected (Fig. 1e), after which the ranking of pathways is available (Fig. 1f). On the results page, the user can access individual metabolic pathways stored on SMPDB [23]. Each node, e.g., metabolite or enzyme, on the website is fully linked to the entry on HMDB [26]. Thus, through this analysis, the user can readily access both SMPDB and HMDB databases (Fig. 2).

Options:

MSEA provides three analytical approaches (Fig. 1b):

(1) Over representation analysis (ORA). This method simply maps the given metabolite list onto the stored metabolite library in the database. This analysis does not consider metabolite concentration data, but can be used for any type of sample to identify which pathways include the user's data.
(2) Single sample profiling (SSP). This analysis requires a list of the metabolite concentration data from a single biofluid sample, e.g., blood and urine. The metabolite concentrations uploaded by the user are compared with the normal range of concentrations reported in the literature. As a typical example, potential metabolite biomarkers showing abnormal concentrations correlated with a particular phenotype can be uploaded, and the specificity of this aberrance among multiple studies can be investigated.
(3) Quantitative enrichment analysis (QEA). This analysis uses a metabolite concentration matrix with either discrete binary or continuous phenotype labels. The rank of enrichment of each pathway is calculated based on the statistical significance of the enrichment.

Depending on the analytical option, the information required as input data is different, e.g., only a list of metabolite names is necessary for ORA. In addition to these analytical approaches, metabolite libraries used for the analysis can be selected (Fig. 1e).

Example Two—Development of a Classification Model Using Metabolite Biomarkers for Discriminating Disease Samples from Controls

The second example addresses the development and validation of a classification model to discriminate patients with a disease from controls, based on observable metabolite patterns. There are often only subtle metabolite changes between two groups, in which case, mathematical models combining multiple metabolites may be useful to improve the ability to discriminate. Here, we use Weka, a freely available software providing a variety of machine learning methods (http://www.cs.waikato.ac.nz/ml/weka/) [33] to build an if-then type decision tree to discriminate metabolite data from humans with/without a disease.

Datasets as inputs: The data matrix used here (Fig. 3a) is slightly modified from the sample file available at the MSEA website (Fig. 1c). The data include metabolite concentrations without any patient ID column. Phenotype labels should be placed in the far right column.

What is available after the analysis?: A classification model discriminating the samples based on their phenotype labels. Their classification accuracies, such as sensitivity and specificity, are also available.

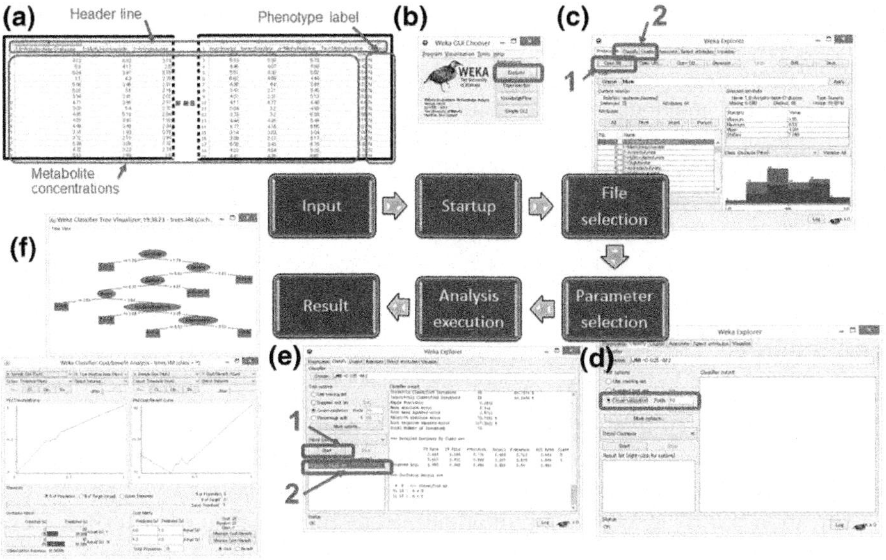

Fig. 3 Analytical flow of Weka software. **a** Data file used for Weka, which is generated by a slight modification of the data for MSEA (Fig. 1a). The first column includes sample information, such as patient ID. The second column includes a phenotype label. For example, labels "Y" and "N" are used for the samples obtained from patients with or without cancer, respectively. The latter columns include metabolite concentrations, the first row of which contains the relevant metabolite name. This file is saved in csv format. **b** Startup panel. The "Explorer" button on the startup pane should be selected. **c** Main panel for data selection. The "Open file" button is selected to load a data file, and the "Classify" tab selected. **d** Main panel for selecting analytical options. After pressing the "Choose" button, "weka-classifiers-trees-J48" is selected, followed by "Cross-validation" in the Test options. **e** The "Start" button (labeled **1**) runs training and validation of the model, and a newly added result can be selected from the "Result list" (labeled **2**). Pressing the right mouse button on this result enables visualization of the results. **f** Decision tree and ROC curves

Overall analytical flow: Through the Weka software, a data file is read, a classification model is selected, and training and evaluation of the model performed (Fig. 3b–c). To evaluate the generalization ability of the developed model, the prediction accuracy is calculated in a cross-validation (CV) manner where a subset of the data is used for model development and the remainder for model evaluation. This process is repeated until all data have been selected to evaluate the model (Fig. 3d, e). The discrimination model is visualized in tree format, including nodes and links, which facilitates understanding the relationship among the metabolites used in the classification model (Fig. 3f). Prediction accuracy is visualized by receiver operating characteristic (ROC) curves to indicate both the sensitivity and specificity of the developed model (Fig. 3f).

Other options: A variety of classification models are implemented on Weka, e.g., support vector machine, artificial neural network, Bayesian network, classification and regression trees (CART), RF, and MLR. Various methods for clustering, feature selection, and resampling of samples are also available. Classification methods that contain a feature selection procedure in their development algorithm, e.g., decision tree, will select only important features for the discrimination. However, other methods, such as MLR, do not select any features and feature selection prior to development of an MLR should be conducted independently. To routinely perform Weka analyses with different options, the calculation program in Weka can be accessed through both an application programmable interface (API) of the Java language, and via command line.

Conclusions

Two commonly used analyses of metabolomics profiles for characterizing human diseases have been introduced. Pathway analysis is used for understanding the enrichment of multiple metabolites categorized into pathways. Development of a classification model incorporating multiple metabolites may be used for the diagnosis and detection of disease. To assess metabolomics profiles, a number of factors are important in addition to analyses such as these, including quality assessment at both a qualitative level (especially for metabolite name assignment) and a quantitative level (by visualizing overall datasets).

References

1. N.L. Kuehnbaum, P. Britz-McKibbin, New advances in separation science for metabolomics: resolving chemical diversity in a post-genomic era. Chem. Rev. **113**(4), 2437–2468 (2013)
2. M.R. Monton, T. Soga, Metabolome analysis by capillary electrophoresis-mass spectrometry. J. Chromatogr. A **1168**(1–2), 237–46 (2007); discussion 236
3. M. Sugimoto et al., Bioinformatics tools for mass spectroscopy-based metabolomic data processing and analysis. Curr. Bioinform. **7**(1), 96–108 (2012)
4. H.C. Keun, T.J. Athersuch, Nuclear magnetic resonance (NMR)-based metabolomics. Methods Mol. Biol. **708**, 321–334 (2011)
5. M. Katajamaa, M. Oresic, Data processing for mass spectrometry-based metabolomics. J. Chromatogr. A **1158**(1–2), 318–328 (2007)
6. M.E. Dumas, J. Kinross, J.K. Nicholson, Metabolic phenotyping and systems biology approaches to understanding metabolic syndrome and fatty liver disease. Gastroenterology **146**(1), 46–62 (2014)
7. S. Shakour et al., Serum metabolomic analysis of pancreatic cancer–letter. Cancer Epidemiol. Biomark. Prev. **22**(10), 1921 (2013)
8. M. Sugimoto et al., Capillary electrophoresis mass spectrometry-based saliva metabolomics identified oral, breast and pancreatic cancer-specific profiles. Metabolomics **6**(1), 78–95 (2010)
9. J.D. Hu et al., Prediction of gastric cancer metastasis through urinary metabolomic investigation using GC/MS. World J. Gastroenterol. **17**(6), 727–734 (2011)
10. J. Xia et al., MetaboAnalyst: a web server for metabolomic data analysis and interpretation. Nucleic Acids Res. **37**(Web Server issue), W652-60 (2009)

11. J. Xia, D.S. Wishart, Web-based inference of biological patterns, functions and pathways from metabolomic data using MetaboAnalyst. Nat. Protoc. **6**(6), 743–760 (2011)
12. N. Kono et al., Pathway projector: web-based zoomable pathway browser using KEGG atlas and Google Maps API. PLoS ONE **4**(11), e7710 (2009)
13. M. Kanehisa et al., KEGG for representation and analysis of molecular networks involving diseases and drugs. Nucleic Acids Res. **38**(Database issue), D355–D360 (2010)
14. H. Rohn et al., VANTED v2: a framework for systems biology applications. BMC Syst. Biol. **6**, 139 (2012)
15. J. Xia, D.S. Wishart, MSEA: a web-based tool to identify biologically meaningful patterns in quantitative metabolomic data. Nucleic Acids Res. **38**(Web Server issue), W71–W77 (2010)
16. A. Subramanian et al., Gene set enrichment analysis: a knowledge-based approach for interpreting genome-wide expression profiles. Proc. Natl. Acad. Sci. U.S.A. **102**(43), 15545–15550 (2005)
17. L. Breiman, Random forests. Mach. Learn. **45**, 5–32 (2001)
18. S.A. Ritchie et al., Metabolic system alterations in pancreatic cancer patient serum: potential for early detection. BMC Cancer **13**, 416 (2013)
19. D.S. Cao et al., A novel kernel Fisher discriminant analysis: constructing informative kernel by decision tree ensemble for metabolomics data analysis. Anal. Chim. Acta **706**(1), 97–104 (2011)
20. T. Kobayashi et al., A novel serum metabolomics-based diagnostic approach to pancreatic cancer. Cancer Epidemiol. Biomark. Prev. **22**(4), 571–579 (2013)
21. T. Soga et al., Serum metabolomics reveals gamma-glutamyl dipeptides as biomarkers for discrimination among different forms of liver disease. J. Hepatol. **55**, 896–905. doi:10.1016/j.jhep.2011.01.031
22. R. Caspi et al., The MetaCyc Database of metabolic pathways and enzymes and the BioCyc collection of Pathway/Genome Databases. Nucleic Acids Res. **36**(Database issue), D623–D631 (2008)
23. A. Frolkis et al., SMPDB: the small molecule pathway database. Nucleic Acids Res. **38** (Database issue), D480–D487 (2010)
24. G. Joshi-Tope et al., Reactome: a knowledgebase of biological pathways. Nucleic Acids Res. **33**(Database issue), D428–D432 (2005)
25. S. Ekins et al., Pathway mapping tools for analysis of high content data. Methods Mol. Biol. **356**, 319–350 (2007)
26. D.S. Wishart et al., HMDB: a knowledgebase for the human metabolome. Nucleic Acids Res. **37**(Database issue), D603–D610 (2009)
27. H. Horai et al., MassBank: a public repository for sharing mass spectral data for life sciences. J. Mass Spectrom. **45**(7), 703–714 (2010)
28. M. Sugimoto et al., MMMDB: mouse multiple tissue metabolome database. Nucleic Acids Res. **40**(Database issue), D809–D814 (2012)
29. Y. Toya et al., 13C-metabolic flux analysis for batch culture of Escherichia coli and its Pyk and Pgi gene knockout mutants based on mass isotopomer distribution of intracellular metabolites. Biotechnol. Prog. **26**(4), 975–992 (2010)
30. Y. Toya et al., Metabolic flux analysis and visualization. J. Proteome Res. **10**(8), 3313–3323 (2011)
31. M. Tomita et al., E-CELL: software environment for whole-cell simulation. Bioinformatics **15** (1), 72–84 (1999)
32. A. Kinoshita et al., Roles of hemoglobin Allostery in hypoxia-induced metabolic alterations in erythrocytes: simulation and its verification by metabolome analysis. J. Biol. Chem. **282**(14), 10731–10741 (2007)
33. E. Frank et al., Data mining in bioinformatics using Weka. Bioinformatics **20**(15), 2479–2481 (2004)

Metagenomics for Monitoring Environmental Biodiversity: Challenges, Progress, and Opportunities

Raghu Chandramohan, Cheng Yang, Yunpeng Cai and May D. Wang

Abstract Metagenomics, as the genomic analysis of DNA materials from environmental samples containing multiple genomic components, is attracting more and more interests due to its wide applications on microbial, cancer, and immunology researches. This chapter provides an overview on the topic covering the major steps involved in data collection, processing, and analysis. We describe and discuss experiment design, sample processing and quality control, sequencing and assembly, annotation, and downstream analyses. For each step, we summarize the current points of views, key issues, and popular tools. A step-by-step tutorial is then given using the popular QIIME pipeline on a bacterial 16S rRNA study case, which would benefit new scientists of the field for the startup of a successful metagenome project.

Introduction

The field of microbial ecology has grown immensely in the past decade with the advent and development of metagenomics. Metagenomics is defined as the direct genetic analysis of genomes contained within an environmental sample [1]. The preliminary projects in the area, involved cloning of environmental DNA, functional expression screening, and then later on lead to the use of shotgun sequencing of these environmental samples' DNA. The technology is a major boon for scientists as it broadened their research focus and gave access to study microbes that

R. Chandramohan · C. Yang · M.D. Wang (✉)
The Joint Department of Biomedical Engineering Department, Georgia Institute of Technology, Emory University, Atlanta, USA
e-mail: maywang@bme.gatech.edu

R. Chandramohan · C. Yang · M.D. Wang
Peking University, Beijing, China

Y. Cai
Research Center for Biomedical Informatics, Shenzhen Institutes of Advanced Technology, Chinese Academy of Sciences, Beijing, China

D. Xu et al. (eds.), *Health Informatics Data Analysis*, Health Information Science,
DOI 10.1007/978-3-319-44981-4_5

cannot be cultured in vitro (only 1% of the microbes can be cultured in vitro) at the sequence level [2]. This innovative approach of studying the microbial population in the environment unveiled the colossal functional gene diversity in microbes.

Metagenomics provides access to the functional gene composition of microbial communities and thus gives a much broader description than phylogenetic surveys, which are often based only on the diversity of one gene, for instance the 16S rRNA gene for bacteria, 18S rRNA or ITS genes for fungus. The data generated from metagenomics is immense and has hastened research in the area of microbial ecology. Some of the applications of metagenomics include identifying novel gene and gene products, genome engineering, understanding cell structure and function, studying the evolution of the genomes in an environmental sample and also elucidating their metabolic network, understanding protein–protein interaction with the help of metaproteomics, and also performing expression studies using metatranscriptomics. With the rapid development of sequencing technology and substantial reduction in cost, metagenomics has become the standard tool in laboratories and scientist working in microbial ecology.

This chapter gives an overview of the field of metagenomics, with particular emphasis on the steps involved in a typical sequence-based metagenome project. We describe and discuss sample processing, sequencing technology, assembly, binning, annotation, experimental design, statistical analysis, and data storage and sharing. This chapter summarizes the current thinking in the field and introduces current practices and key issues that those scientists new to the field need to consider for a successful metagenome project.

Sampling and Experimental Design

Earlier computational genomics project was more exploratory and did give the due importance to the reproducibility of the experiment. But as we go forward in this era, we see the rising importance of proving our experiments with statistical rigor. Likewise in metagenomics, we need to perform sampling in such a way that our results can be reproduced and are statistically significant. Before beginning an experiment, a power analysis must be performed to decide the number of samples required to perform a meaningful analysis. The procedure we employ to sample is also just as important as the number of samples we need for our experiment. Since we are dealing with environmental samples, we must not bias the DNA we are collecting. The sample collected must have accurate quantities representing the true composition of the microbial environment. The bias can be reduced by taking replicates of sample as we do not initially know which organisms we are trying to capture. It should not be that we take a sample and separate it into aliquots before sequencing to get replicates, i.e., they must be picked individually. Also, using extremely high depth/coverage used in sequencing does not correlate with increased accuracy of the true observation. When considering environmental sample, it is also critical to consider temporal variations. Since the environment changes over time, the sampling for the experiment must be done in a preferably similar time frame to avoid factors involved with time. If we are to perform sampling on a host, the primary problem we observe here is that the host DNA overwhelms the microbial DNA, especially if it is large.

Typical factors that should be considered in metagenomic experiments include the choice of sampling spots (dimension of sample materials, number of samples, and distribution of spots), sampling time, sequencing type, and sequencing depth. Metagenomic objects can be very sensitive to slight fluctuations of environmental factors. For example, in soil or human microbe sampling, a small shift of sampling position may result in dramatic changes in microbe components. Hence, a clear and rigid protocol for choosing sampling spots is necessary in order to properly reflect the goal of the experiment. On the other hand, the biodiversity of the metagenomic objects under investigation can be often far above the expectation of the researchers, and the key components of the object which carry the desired property may take up only a small portion of the biological community. A poor designed sequencing scheme will lead to inadequate precision on profiling the key components, which may be disastrous to the entire experiment. To alleviate this situation, a wise strategy is to carry out a pilot study [3] for acquiring knowledge about the biodiversity and variations of the environmental materials, and revise the experiment design according to the outcome of the pilot data.

Prior statistical tests developed in quantitative ecology of higher organisms were among the first to be employed for metagenomic data analyses such as SIMPER and ANOSIM. Statisticians have since then come up with statistical tests which give an insight into the core composition of the metagenomic sample and to perform other statistical analysis. As metagenomic data does not follow a normal distribution and has a long skewed tail, we cannot employ directly many of the parametric tests. Thus, statisticians continuously develop nonparametric tests through simulations, resampling, or permutations that fit the data better.

There have been many experiments in the past which have been poorly designed and whose results could not be reproduced. The area of metagenomics is exciting to statisticians in that each experiment is unique and designing a meaningful test is much appreciated.

Metagenomic Sequencing and Preprocessing

Sequencing technologies employed in metagenomic studies can be generally classified into three categories: whole genome shotgun (WGS) sequencing, amplicon sequencing, and transcriptome sequencing. Whole genome shotgun sequencing explores the entire gene material without any predefined filter, resulting in a survey on the distribution of gene components. Amplicon sequencing, on the other hand, focuses on a small specific region on the genome starting with a conservative tag sequence (called primer) to achieve detailed knowledge of the region with ultra high sequencing coverage. Amplicon sequencing has been a powerful tool for in-depth analysis of phylogenetic and evolutionary details. Transcriptome sequencing, also called RNA-seq, adopts a RNA-specific primer to extract and clone the message RNA components of the sampled material, and perform shotgun sequencing resulting in a quantitative spectrum of the RNA transcription activities. Researchers typically discriminate the usages of these sequencing techniques according to the following purposes [4]: "Who are they?"—for amplicon sequencing; "What can they do?"—for whole genome sequencing; and "What are they doing?"—for transcriptome sequencing. However, this distinction is not rigid. For example, by using many samples, species tag (e.g., 16S rRNA) amplicon sequencing can also be adopted to model the interactions between microbial species in the nature environments.

Preprocessing of metagenomic sequencing data usually includes three basic steps: quality control, removing barcodes, and removing chimeric sequences.

- Quality control: Nucleotides with low quality scores should be treated unreliable and marked as an ambiguous nucleotide, or removed for the sequence. On the other hand, if a sequence does not contain a continuous high quality segment with sufficient length (the threshold should be determined by the researcher himself), the entire sequence should be abandoned.
- Removing barcodes: Currently, in order to save sequencing cost, most metagenomics projects employ barcode techniques to sequence multiple samples in the same run. A barcode is a predesigned short sequence attached to one end of a sequence to identify its source. Barcode is not part of the original sequence and should be removed before analyses, which can be done using simple program codes.
- Chimera removal: Chimera is a phenomenon that happens in the library preparing phase of sequencing and causes structural error in sequencing. Chimera happens when a DNA copying is interrupted midway and the partly copied segment is detached from the source A, then reattached to a difference source B, resulting in a chimeric sequence with one part analogue with A and others with B. Because metagenomic sequencing, especially amplicon sequencing, involves lots of similar sequence components, the impact of chimera is considerable. Currently, there is not golden standard for identifying chimeric sequences. Nevertheless, a number of tools do exist to help identifying sequences that are potentially chimeric.

Metagenomic Data Analyses

a. **Identification of Metagenomics Community Composition**

The first step in metagenomics study is typically to provide a detailed description of the population composition about studied sample. This can be generally carried out from two aspects: the ecological scope and the functional scope. In this subsection, we will introduce the idea of ecological analysis before talking about functional analysis in the next subsection (metagenomics annotation). The basic task of ecological study is to identify the taxonomic compositions of the sampled material and their abundances from the sequence data, which is called sequence binning. Existing technologies of metagenomic sequence binning can be classified into two categories: taxonomy-dependent approaches, which match sequences to a reference library to determine their sources, and taxonomy-independent approaches, which adopt clustering technology to group the sequences into operational taxonomy units (OTUs) based on their internal similarities without external references. The most frequently used technique for taxonomy-dependent analysis is to adopt sequence alignment tools such as BLAST [5] or BLAT [6] to find a most similar reference for each sequence, and use the annotation of the reference as the taxonomy assignment. More advanced techniques, such as CARMA [7], AMPHORA [8], and INFERER [9], adopt probabilistic models such as Hidden Markov models to reflect the sequence characters of a reference taxon with intra-taxon variation considered, and align query sequences against the model, which usually achieves better assignment but with high computational burden. Due to the nature of metagenomics that unknown species may take up a significant portion of the studied community, taxonomy-independent analysis is often preferred in providing a complete description of the biological community. Dozens of methods have been proposed for this purpose and the underlying techniques are quite diverse. Currently, TETRA [10], CompostBin [11], and MetaCluster [12] are leading methods for taxonomy-independent binning of whole genome or transcription sequences, while USEARCH [13] and ESPRIT-Tree [14] are currently two most up-to-date algorithms for amplicon sequences that can handle millions of reads.

b. **Metagenomics Annotation**

Metagenomic annotation usually consists of two steps: (1) ORF (open reading frame) calling or gene prediction from contigs and (2) functional annotation of ORFs or genes. Gene prediction can be categorized into two groups, including (1) evidence-based methods relying on homology search, which can find only previously known genes; and (2) ab initio algorithms, which are capable of detecting novel genes [15]. For evidence-based metagenomic gene prediction methods, the selections can be comparisons against known protein databases with BLAST packages, CRITICA [16], and Orpheus [17]. As for ab initio algorithms, considerable methods have been designed specifically for metagenomic DNA fragments such as MetaGeneMark [18], MetaProdigal [19], Glimmer-MG [20], Orphelia [21], MetaGUN [15], and FragGeneScan [22]. Most algorithms use Hidden Markov models and machine learning approaches to approximate optimized parameters from training sets for prediction. The ab initio algorithms show reliable performance (accuracy around 97%) on the long DNA fragments (>500 bp), while the performance (accuracy around 89%) declines severely for short fragments (<120 bp) [15], indicating it necessary to proceed assembly for the NGS sequence. Note that the development in NGS techniques and metagenomic assembly have already been able to produce long enough contigs [23, 24], enhancing the feasibility of gene prediction.

For most metagenomic projects, one of the major computation challenges is function annotation [25]. Annotation is typically done with homology searches by mapping to gene or protein databases, such as COG/KOG [26], eggNOG [27], KEGG [28], PFAM [29], and TIGRFAM [30]. But as estimated, less than 50% of a metagenomic sequence can be annotated [31]. Until now, no reference database can cover all biological functions. Therefore, visualizing and merging the interpretations of all database searches within a single framework is an essential task [25], as implemented in several metagenomic analysis platforms that will be introduced in later sections.

c. **Downstream Analyses**

By sequence binning and annotation, the taxonomy and function compositions of the metagenomics have been identified, which enable downstream analyses of the sequence data. Typically, metagenomics sequences are reduced into a summary table with row representing the operational taxonomic unit (OTU) and the columns representing the category (organism classification, gene function classification, enzyme classification, pathways classification), on which many statistical tests that can handle this kind of data.

Ecological statistics is one of the most frequently used analyses that provide a full picture of the metagenomic community from sequencing results. The key task of ecological statistics is to infer the taxon diversity of the environment from the samples. Because biological communities usually follow a long-tail distribution and the samples cannot guarantee to cover all taxa due to limited sample depth, statistical inferences should be employed to estimate the number of potential taxa. The alpha diversity describes the estimated taxon diversity of an environment sample inferred from sequencing results. Typical statistics indices include the CHAO1 estimation [32], the ACE estimation [33], and the rarefaction curve [34]. The beta diversity, on the other hand, measures the differentiations among samples. Simple metrics of beta diversity include Jaccard index, Sorensen index, Simple Matching Coefficient, etc. These indices are too rough to reflect the quantitative variation between samples. The Unifrac distance [35] was proposed specifically for evaluating the inter-sample dissimilarities in metagenomics, which measures the similarity between two samples by merging their data together and performing clustering.

In contrast to statistical inferences, phylogenetic analysis studies the relationship between individual taxa within a sample or among different samples and reveals their evolutionary relationship based on sequence similarities, in the form of a phylogenetic (evolutionary) tree. Currently, practical approaches of constructing phylogenetic trees can be classified into two categories: minimal evolution ones (NINJA [36], QuickTree [37], FASTTree [38]), which minimize the overall span of the phylogenetic tree, and maximum likelihood ones (PhyML [39], RAxML [40], FASTTree-2 [41]), which find an evolutionary relationship that maximizes the probability of generating the observed data.

Principle coordinate analysis (PCoA) and interaction networks are another two frequently used techniques for in-depth studies of the interactions between the metagenomics community and the environment or among community components. Matching the principle coordinates of variation to environment factors has led to various important metagenomic discoveries such as the entero-types of human gut microbe [42]. On the other hand, interaction networks are usually the starting point for describing how the influence of environments propagates within the studied community and what kind of compositions play important roles in it.

d. **Data Visualization**

"Communicating information clearly and effectively" is the main goal of data visualization [43], which has become an essential part of metagenomic analysis. Some metagenomics analysis platforms, such as IMG/M [44] and MEGAN [45], have integrated visualization tools. However, as metagenomic analysis platforms must consider other aspects of analysis, which might limit their performance on visualization. Here, we confine to the tools designed specifically for metagenomic visualization, such as Krona [46], MetaSee [47], and VAMPS [48]. Most metagenomic visualization tools focus on the phylogenetic annotation [49] and the assignment of taxonomy or gene function [46, 48]. With the visualization of phylogenetic information and taxonomic or functional hierarchies, researchers can obtain an overview of the microbial population structures and articulate the difference of samples. However, metagenomic samples comprise numerous known or unknown species, and this uncertainty increases the "levels of granularity inherent in these classifications" [46]. To address this challenge, Krona [46] proposed to use radial space-filling displays that can illustrate hierarchical data with zoomable pie charts in an intuitive interactive way. Even though Krona can display the structure of single sample vividly, displaying only one sample in a window at a time limits the convenience of comparing samples [47]. Therefore, MetaSee [47] presented a toolbox, which includes a critical part that integrates various views for the comparison of multiple samples, for metagenomic visualization. As for recently developed VAMPS [48] and Amphora Vizu [49], they both allow using marker genes to analyze the diversity of microbial communities and phylogenetic classification, respectively. Apart from these aforementioned visualization tools for metagenomic, many other tools, such as Explicet [50] and SynTView [51], are still waiting to be explored by researchers. Facilitated by the next generation sequencing technologies, a large amount of available metagenomic data require more efficient, interactive, and extendable visualization.

Platforms

IMG/M is an integrated metagenome data management and comparative analysis system which integrates metagenome data sets with isolate microbial genomes from IMG [44, 52]. It integrates all datasets into a single protein level abstraction [25]. As for MG-RAST, it provides the functions of data repository, analysis pipeline, and genomics comparison. "It has been optimized for achieving a trade-off between accuracy and computational efficiency for short reads" [25]. Until March 2014, MG-RAST has over 12,000 registered users, 110,593 data sets, 110,593 metagenomes, and 43.74 Terabase. These statistics demonstrate MG-RAST's representative status in the standardize pipeline. "CAMERA [53] offer more flexible annotation schema but require that individual researchers understand the annotation of data and analytical pipelines well enough to be confident in their interpretation" [25]. Recently EBI metagenomics—a new resource for the analysis and archiving of metagenomics data [54] was lunched; it is the first metagenomics analysis pipeline in Europe, which also undergoes quality control checks, and functional and taxonomic analyses. All the above resources (IMG/M, MG-RAST, CAMERA, EBI metagenomics) "have representatives in the Genomic Standards Consortium (GSC) [55] and have all adopted and implemented the MIxS (minimum information about any sequence) [56] checklists" [54]. As the state-of-the-art resources, all the platforms share almost the same workflow, and the following effort is to "establish a platform for next generation collaborative computational infrastructures, called M5 (Metagenomics, Metadata, MetaAnalysis, Models and MetaInfrastructure) in which all parties are involved" [54, 57]. With multiple choices, no doubt it will be more easily for users to access and interpret the data.

For metagenomics analyses of bacteria 16S rRNA sequences, established pipelines include QIIME [58], mother [59], and RDP-pyro [60].

Example: Bacteria 16S RRNA Metagenomics Pipeline

In this section, we provide a simple example of analyzing bacteria 16S rRNA metagenomics sequences using popular bioinformatic tools. Suppose you are going to study the differences of gut bacteria composition between patient and healthy people, and you choose the V4 region of the 16S rRNA as the sequencing target. The first step is to design a sequencing scheme as below:

```
#SampleID BarcodeSequence LinkerPrimerSequence Treatment ReversePrimer
Sample1 CGCTTATCGAGA GTGTGCCAGCMGCCGCGGTAA patient GGACTACHVGGGTWTCTAAT
Sample2 CATACCAGTAGC GTGTGCCAGCMGCCGCGGTAA patient GGACTACHVGGGTWTCTAAT
Sample3 CTCTCTACCTGT GTGTGCCAGCMGCCGCGGTAA healthy GGACTACHVGGGTWTCTAAT
........................
........................
```

The above scheme adopts the 515F/806R primer pair to capture the V4 hypervariation region, and assign a barcode sequence to each sample. Moreover, a two-base linker 'GT' is attached to the forward primer to ensure the quality of sequencing the primer and the barcode. The choice of barcodes and linkers is device-specific and should consult the device guideline for optimal solutions. Note that although the reverse primer is not always needed for sequencing, it is preferred for helping with validating the integrity of the sequences obtained. We record the scheme in a text file 'samplemapping.txt'.

Step 1: Preprocessing

Suppose you used a 454 GS FLX sequencer, the sequencing results are composed of two major files: a '.fna' file containing the sequences in FASTA format and a '. qual' file containing the quality score of each nucleotide base. We process the data in the QIIME environment using the following command:

```
$ split_libraries.py -m samplemapping.txt -f samples.fna -q
samples.qual -z truncate_remove -w 50 -g -o preproc
```

The above command produces an output file `preproc/seqs.fna` which contains filtered sequences. The barcodes and primers are removed from the sequences and the sample name is attached to the sequence label. Moreover, sequences with insufficient length or poor quality sores are excluded from the data. We further perform chimera removal:

```
$ identify_chimeric_seqs.py -i preproc/seqs.fna -m use-
arch61 -o chimeras/-r reflib.fna
$ filter_fasta.py -f preproc/seqs.fna -o seqs_filtered.fna
-s chimeras/chimeras.txt -n
```

Here '`reflib.fna`' is a reference library of reliable bacteria 16S rRNA sequences. A set of libraries have been provided by QIIME and other tools.

Step 2: OTU picking and annotation

The following command generates operational taxonomy units (OTUs) using USEARCH under the QIIME environment:

```
$ pick_de_novo_otus.py -m usearch61 -i seqs_filtered.fna -
o otus
```

The above command generates a file with suffix '`_otu.txt`' in the `otus/uclust_picked_otus` directory which contains the sequence labels of each taxonomy unit per line. Moreover, for each OUT, a representative sequence is generated in `otus/rep_set directory,` along with their taxonomy annotations. The presentative sequences are further aligned and a phylogenetic tree is generated, which is required for downstream analyses. Finally, a taxonomy summary table about the distribution of microbial compositions in different samples is given in the file `otus/otu_table.biom`, which is a JSON file in BIOM format.

Step 3: Visualization and downstream analyses

The following command parses the taxonomy table generated in the above step and generates a series of figures describing the distributions of known taxonomy components in different biological taxonomy levels:

```
$ summarize_taxa_through_plots.py -i otus/otu_table.
biom -o taxa_summary -m samplemapping.txt
```

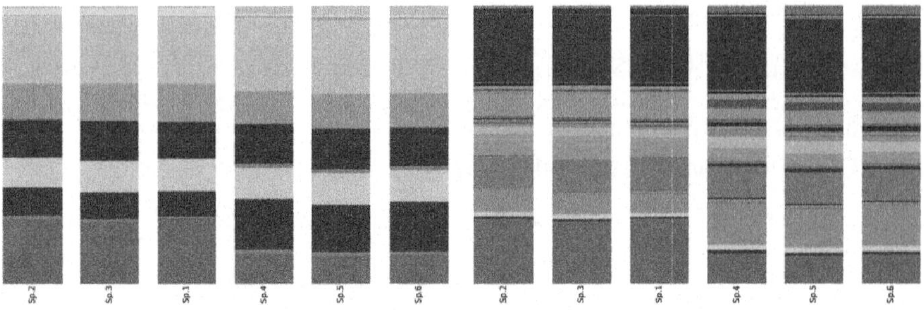

Fig. 1 Example of a taxonomy distribution graph on different levels: (*left*) phylum, (*right*) family

Figure 1 depicts the results of a taxonomy distribution graph on phylum and family levels, respectively. The results are stored in .html format under the `/taxa_summary` directory.

The following command provides more detailed information about taxonomy compositions among samples in the form of a heat map. The results are stored in a PDF format. Figure 2 depicts an example of the execution results.

```
$ make_otu_heatmap.py -i taxa_summary/otu_table_L4.biom
-o taxa_summary
```

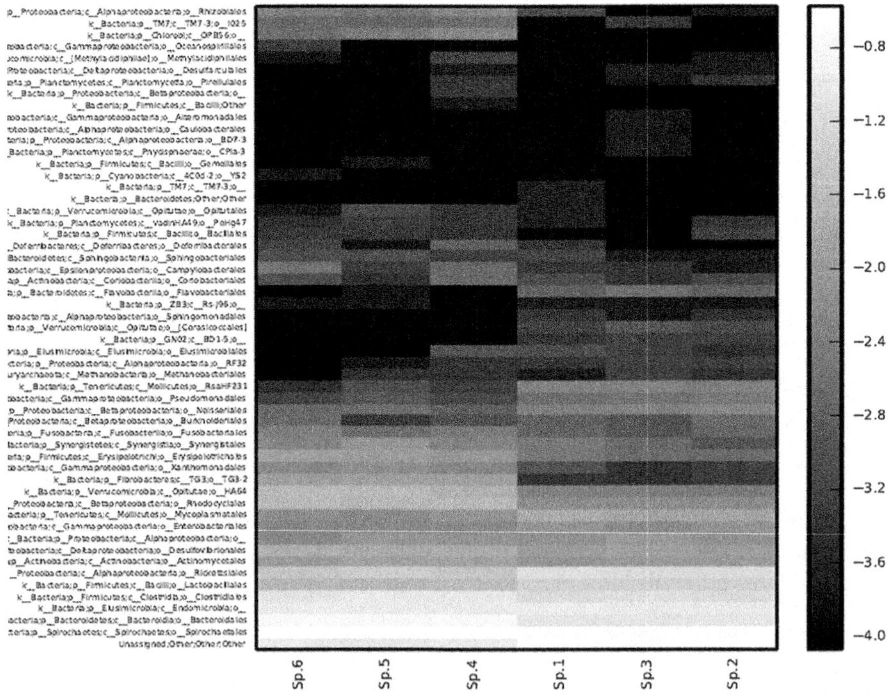

Fig. 2 Example of a taxonomy heatmap on family level

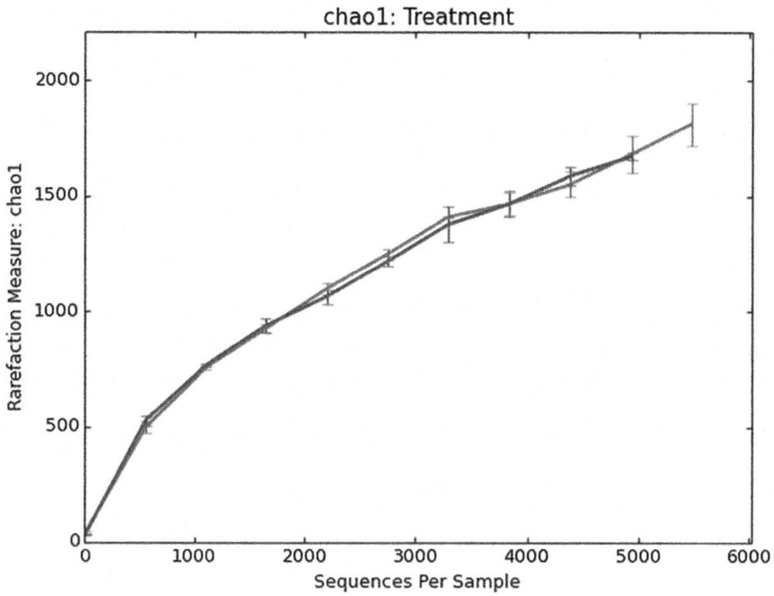

Fig. 3 Example of a rarefaction curve plot, *blue* healthy group, *red* patient group (color figure online)

The following command analyzes the alpha diversity of each sample and plots a set of rarefaction curves with CHAO1 estimation on various sample depths. Figure 3 depicts an example of the obtained rarefaction curve plots. The results figures are stored in the /alpha directory.

```
$ alpha_rarefaction.py -p param.txt -i otu_table.biom -m samplemapping.txt -o alpha
```

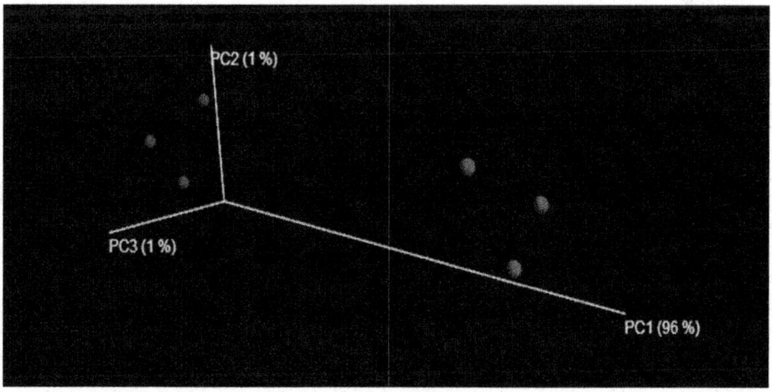

Fig. 4 Example of principle coordinate analysis to describe the similarity and differences between sampled environments. *Blue* healthy, *Red* patients. We see the two groups are separated clearly (color figure online)

The following command analyzes the beta diversity between samples and performs principle coordinate analysis (PCoA) to visualize the differences. Figure 4 depict the results of PCoA. The results figures are stored in the /beta directory.

```
$ beta_diversity_through_plots.py -i otu_table.biom -m
samplemapping.txt -o beta
```

Conclusion

Genome sequencing is now a common tool in biological and biomedical research. In the past few years, the data generation capacity of high-throughput sequencing has increased dramatically at a speed exceeding the Moore's law and with sharply reduced cost. The extensive accumulation of genomic information represents a valuable source for expanding biological knowledge. As a result, metagenomics has become an explosive increasing topic for dramatically changing the traditional paradigm of ecological studies. Bioinformatics plays an important role in this area for providing powerful tools to handle the massive amount of data. As introduced in this section, dozens of tools have been proposed to aid various steps of metagenomic analysis covering from preprocessing to biological statistics. Nevertheless, it should be pointed out that existing tools are still far from optimal in accurately detecting the composition of the metagenomics communities, even with very simple cases. Serious challenges persist regarding the accuracy and computational speed of existing data processing pipelines. Hence, developing of efficient bioinformatic methods for fast and accurate analyses of huge and growing sequencing data will be an essential task in the future development of metagenomics.

References

1. T. Thomas, J. Gilbert, F. Meyer, Metagenomics—a guide from sampling to data analysis. Microb. Inform. Exp. **2**, 3 (2012)
2. R.I. Amann, B.J. Binder, R.J. Olson, S.W. Chisholm, R. Devereux, D.A. Stahl, Combination of 16S rRNA-targeted oligonucleotide probes with flow cytometry for analyzing mixed microbial populations. Appl. Environ. Microbiol. **56**, 1919–1925 (1990)
3. J. Handelsman, J. Tiedje, L. Alvarez-Cohen et al., The new science of metagenomics: revealing the secrets of our microbial planet. Nat. Res. Counc. Rep. **13**, 60–65 (2007)
4. J.M.D. Bella, Y. Bao, G.B. Gloor, J.P. Burton, G. Rrid, High throughput sequencing methods and analysis for microbiome research. J. Microbiol. Methods **95**, 401–414 (2013)
5. S.F. Altschul, T.L. Madden, A.A. Schaffer et al., Gapped BLAST and PSI-BLAST: a new generation of protein database search programs. Nucl. Acids Res. **25**(17), 3389–3402 (1997)
6. K.W. James, BLAT—the BLAST-like alignment tool. Genome Res. **12**(4), 656–664 (2002)
7. L. Krause, N.N. Diaz, A. Goesmann et al., Phylogenetic classification of short environmental DNA fragments. Nucleic Acids Res. **36**(7), 2230–2239 (2008)
8. M. Wu, J.A. Eisen, A simple, fast, and accurate method of phylogenomic inference. Genome Biol. **9**(10), R151 (2008)

9. E.P. Nawrocki, L.K. Diana, L. Kolbe, S.R. Eddy, Infernal 1.0: inference of RNA alignments. Bioinformatics **25**(10), 1335–1337 (2009)
10. H. Teeling, J. Waldmann, T. Lombardot et al., TETRA: a web-service and a stand-alone program for the analysis and comparison of tetranucleotide usage patterns in DNA sequences. BMC Bioinformatics **5**, 163 (2004)
11. S. Chatterji, I. Yamazaki, Z. Bai, et.al., CompostBin: a DNA composition-based algorithm for binning environmental shotgun reads, in *Research in Computational Molecular Biology* (Springer, Berlin, 2008), pp. 17–28
12. H.C.M. Leung, S.M. Yiu, B. Yang et al., A robust and accurate binning algorithm for metagenomic sequences with arbitrary species abundance ratio. Bioinformatics **27**(11), 1489–1495 (2011)
13. R.C. Edgar, Search and clustering orders of magnitude faster than BLAST. Bioinformatics **26** (19), 2460–2461 (2010)
14. Y. Cai, Y. Sun, ESPRIT-Tree: hierarchical clustering analysis of millions of 16S rRNA pyrosequences in quasilinear computational time. Nucleic Acids Res. **39**(14), e95 (2011)
15. Y. Liu, J. Guo, G. Hu, H. Zhu, Gene prediction in metagenomic fragments based on the SVM algorithm. BMC Bioinformatics **14**, S12 (2013)
16. J.H. Badger, G.J. Olsen, CRITICA: coding region identification tool invoking comparative analysis. Mol. Biol. Evol. **16**, 512–524 (1999)
17. D. Frishman, A. Mironov, H.-W. Mewes, M. Gelfand, Combining diverse evidence for gene recognition in completely sequenced bacterial genomes. Nucleic Acids Res. **26**, 2941–2947 (1998)
18. W. Zhu, A. Lomsadze, M. Borodovsky, Ab initio gene identification in metagenomic sequences. Nucleic Acids Res. **38**, e132–e132 (2010)
19. D. Hyatt, P.F. LoCascio, L.J. Hauser, E.C. Uberbacher, Gene and translation initiation site prediction in metagenomic sequences. Bioinformatics **28**, 2223–2230 (2012)
20. D.R. Kelley, B. Liu, A.L. Delcher, M. Pop, S.L. Salzberg, Gene prediction with Glimmer for metagenomic sequences augmented by classification and clustering. Nucleic Acids Res. **40**, e9 (2012)
21. K.J. Hoff, M. Tech, T. Lingner, R. Daniel, B. Morgenstern, P. Meinicke, Gene prediction in metagenomic fragments: a large scale machine learning approach. BMC Bioinformatics **9**, 217 (2008)
22. M. Rho, H. Tang, Y. Ye, FragGeneScan: predicting genes in short and error-prone reads. Nucleic Acids Res. **38**, e191–e191 (2010)
23. J. Qin, R. Li, J. Raes, M. Arumugam, K.S. Burgdorf, C. Manichanh et al., A human gut microbial gene catalogue established by metagenomic sequencing. Nature **464**, 59–65 (2010)
24. T. Namiki, T. Hachiya, H. Tanaka, Y. Sakakibara, MetaVelvet: an extension of Velvet assembler to de novo metagenome assembly from short sequence reads. Nucleic Acids Res. **40**, e155–e155 (2012)
25. T. Thomas, J. Gilbert, F. Meyer, Metagenomics—a guide from sampling to data analysis. Microb. Inform. Exp. **2** (2012)
26. R.L. Tatusov, N.D. Fedorova, J.D. Jackson, A.R. Jacobs, B. Kiryutin, E.V. Koonin et al., The COG database: an updated version includes eukaryotes. BMC Bioinformatics **4**, 41 (2003)
27. J. Muller, D. Szklarczyk, P. Julien, I. Letunic, A. Roth, M. Kuhn et al., eggNOG v2. 0: extending the evolutionary genealogy of genes with enhanced non-supervised orthologous groups, species and functional annotations. Nucleic Acids Res. **38**, D190–D195 (2010)
28. M. Kanehisa, S. Goto, S. Kawashima, Y. Okuno, M. Hattori, The KEGG resource for deciphering the genome. Nucleic Acids Res. **32**, D277–D280 (2004)
29. M. Punta, P.C. Coggill, R.Y. Eberhardt, J. Mistry, J. Tate, C. Boursnell et al., The Pfam protein families database. Nucleic Acids Res. **40**, D290–D301 (2012)
30. J.D. Selengut, D.H. Haft, T. Davidsen, A. Ganapathy, M. Gwinn-Giglio, W.C. Nelson et al., TIGRFAMs and Genome Properties: tools for the assignment of molecular function and biological process in prokaryotic genomes. Nucleic Acids Res. **35**, D260–D264 (2007)

31. J.A. Gilbert, D. Field, P. Swift, S. Thomas, D. Cummings, B. Temperton et al., The taxonomic and functional diversity of microbes at a temperate coastal site: a 'multi-omic' study of seasonal and diel temporal variation. PLoS ONE **5**, e15545 (2010)
32. A. Chao, Non-parametric estimation of the number of classes in a population. Scand. J. Stat. **11**, 265–270 (1984)
33. A. Chao, S.M. Lee, Estimating the number of classes via sample coverage. J. Am. Stat. Assoc. **87**, 210–217 (1992)
34. S.H. Hurlbert, The non-concept of species diversity: a critique and alternative parameters. Ecology **52**, 577–586 (1971)
35. C. Lozupone, R. Knight, UniFrac: a new phylogenetic method for comparing microbial communities. Appl. Environ. Microbiol. **71**(12), 8228–8235 (2005)
36. T.J. Wheeler, Large-scale neighbor-joining with NINJA, in *Algorithms in Bioinformatics* (Springer, Berlin, 2009), pp. 375–389
37. K. Howe, A. Bateman, R. Durbin, QuickTree: building huge Neighbour-Joining trees of protein sequences. Bioinformatics **18**(11), 1546–1547 (2002)
38. M.N. Price, P.S. Dehal, A.P. Arkin, FastTree: computing large minimum evolution trees with profiles instead of a distance matrix. Mol. Biol. Evol. **26**(7), 1641–1650 (2009)
39. S. Guindon, et al., New algorithms and methods to estimate maximum-likelihood phylogenies: assessing the performance of PhyML 3.0. Syst. Biol. **59**(3), 307–321 (2010)
40. Alexandros Stamatakis, RAxML-VI-HPC: maximum likelihood-based phylogenetic analyses with thousands of taxa and mixed models. Bioinformatics **22**(21), 2688–2690 (2006)
41. M.N. Price, P.S. Dehal, A.P. Arkin, FastTree 2—approximately maximum-likelihood trees for large alignments. PLoS ONE **5**(3), e9490 (2010)
42. M. Arumugam et al., Enterotypes of the human gut microbiome. Nature **473**(7346), 174–180 (2011)
43. V. Friedman, Data visualization and infographics. Graph. Monday Inspiration **14**, 2008 (2008)
44. V.M. Markowitz, I.-M.A. Chen, K. Chu, E. Szeto, K. Palaniappan, Y. Grechkin et al., IMG/M: the integrated metagenome data management and comparative analysis system. Nucleic Acids Res. **40**, D123–D129 (2012)
45. D.H. Huson, S. Mitra, H.-J. Ruscheweyh, N. Weber, S.C. Schuster, Integrative analysis of environmental sequences using MEGAN4. Genome Res. **21**, 1552–1560 (2011)
46. B.D. Ondov, N.H. Bergman, A.M. Phillippy, Interactive metagenomic visualization in a Web browser. BMC Bioinformatics **12**, 385 (2011)
47. B. Song, X. Su, J. Xu, K. Ning, MetaSee: an interactive and extendable visualization toolbox for metagenomic sample analysis and comparison. PLoS ONE **7**, e48998 (2012)
48. S.M. Huse, D.B.M. Welch, A. Voorhis, A. Shipunova, H.G. Morrison, A.M. Eren et al., VAMPS: a website for visualization and analysis of microbial population structures. BMC Bioinformatics **15**, 41 (2014)
49. C. Kerepesi, B. Szalkai, V. Grolmusz, Visual analysis of the quantitative composition of metagenomic communities: the AmphoraVizu Webserver. Microb. Ecol. 1–3 (2014)
50. C.E. Robertson, J.K. Harris, B.D. Wagner, D. Granger, K. Browne, B. Tatem, et al., Explicet: Graphical user interface software for metadata-driven management, analysis, and visualization of microbiome data. Bioinformatics btt526 (2013)
51. P. Lechat, E. Souche, I. Moszer, SynTView—an interactive multi-view genome browser for next-generation comparative microorganism genomics. BMC Bioinformatics **14**, 277 (2013)
52. S. Möller, M.D. Croning, R. Apweiler, Evaluation of methods for the prediction of membrane spanning regions. Bioinformatics **17**, 646–653 (2001)
53. S. Sun, J. Chen, W. Li, I. Altintas, A. Lin, S. Peltier et al., Community cyberinfrastructure for advanced microbial ecology research and analysis: the CAMERA resource. Nucleic Acids Res. **39**, D546–D551 (2011)
54. S. Hunter, M. Corbett, H. Denise, M. Fraser, A. Gonzalez-Beltran, C. Hunter et al., EBI metagenomics—a new resource for the analysis and archiving of metagenomic data. Nucleic Acids Res. **42**, D600–D606 (2014)

55. D. Field, L. Amaral-Zettler, G. Cochrane, J.R. Cole, P. Dawyndt, G.M. Garrity, et al., The genomic standards consortium. PLoS Biol. **9** (2011)
56. P. Yilmaz, R. Kottmann, D. Field, R. Knight, J.R. Cole, L. Amaral-Zettler et al., Minimum information about a marker gene sequence (MIMARKS) and minimum information about any (x) sequence (MIxS) specifications. Nat. Biotechnol. **29**, 415–420 (2011)
57. E. Glass, F. Meyer, J.A. Gilbert, D. Field, S. Hunter, R. Kottmann et al., Meeting report from the genomic standards consortium (GSC) workshop 10. Stand. Genomic Sci. **3**, 225 (2010)
58. J.G. Caporaso, J. Kuczynski, J. Stombaugh et al., QIIME allows analysis of high-throughput community sequencing data. Nat. Methods **7**(5), 335–336 (2010)
59. P.D. Schloss, S.L. Westcott, T. Ryabin et al., Introducing mothur: open-source, platform-independent, community-supported software for describing and comparing microbial communities. Appl. Environ. Microbiol. **75**(23), 7537–7541 (2009)
60. J.R. Cole, Q. Wang, J.A. Fish et al., Ribosomal Database Project: data and tools for high throughput rRNA analysis. Nucl. Acids Res. **41**, D633–D642 (2014)

Clinical Assessment of Disease Risk Factors Using SNP Data and Bayesian Methods

Ivan Kozyryev and Jing Zhang

Abstract Recent groundbreaking technological and scientific achievements impelled the field of personalized medicine (PM), which promises to start a new era in clinical disease treatment. However, the degree of success of PM strongly depends on the establishment of a vast resource library containing the connections between many common complex diseases and specific genetic signatures. Particularly, these connections can be discovered performing whole-genome association studies, which attempt to link diseases to their genetic origins. Such large-scale surveys, combined with modern advanced statistical methods, have already identified many disease-related genetic variants. In this review, we describe in detail novel statistical methods based on Bayesian data analysis ideas—Bayesian modeling, Bayesian variable partitioning, and Bayesian graphs and networks—which are promising to help shine light on complex biological processes involved in disease formation and development. Particularly, we outline how to use Bayesian approaches in the context of clinical applications to perform epistasis analysis while accounting for the block-type genome structure.

Promise and Complexity of Personalized Medicine

Simple and inexpensive genetic tests capable of showing person's risks to develop certain diseases would help to effectively target clinical treatments to each individual patient in order to achieve the best possible results [1, 2]. Consequently, efficient technologies and software for uncovering treatment-related mutations in

I. Kozyryev
Department of Physics, Harvard University, Cambridge, MA, USA

J. Zhang (✉)
Department of Mathematics and Statistics, Georgia State University, Atlanta, GA, USA
e-mail: jzhang47@gsu.edu

D. Xu et al. (eds.), *Health Informatics Data Analysis*, Health Information Science,
DOI 10.1007/978-3-319-44981-4_6

illness-inducing viruses as well as disease-related variants in patient's DNA will play important roles in the future of medicine. Improved disease prevention and diagnosis as well as novel routes to therapies are the main motivations for extensive studies aimed at finding disease-related genetic signatures.

Presently, the estimated disease risks via characterization of known genetic risk factors can provide only a limited help in clinical applications [1, 3]. Even though a large amount of resources has been directed in this direction recently, the genetic basis of common human diseases has not been identified for the most part [4, 5]. Recent emergence of successful experimental and statistical strategies for the genome-wide association studies was supposed to provide the necessary tools for deciphering genetic causes of complex human illnesses like type 1 and 2 diabetes [6], rheumatoid arthritis, and bipolar disorder [4, 7]. However, the presence of complicated multi-locus interactions immensely complicates the task of discovering disease-related variants in patient's genome [8, 9]. Thus, biochemical and statistical understanding of genetic interactions will play a crucial role in future clinical applications.

Whole-Genome Association Studies

An examination of a large number of genetic markers across the whole genome for multiple individuals with the goal of identifying variants-disease associations is known as genome-wide association study (GWAS). Novel scientific and technological advances in high-throughput biotechnologies such as microarrays and next-generation sequencing [10–12] made GWAS a powerful tool for unlocking the genetic basis of complex diseases. Particularly, development of International HapMap resource [13] that simplified design and analysis of association studies, emergence of dense genotyping chips [10, 14], and assembly of large and characterized clinical samples [4] should be singled out as important factors in recent successful progress for GWAS. While many disease loci have been identified in such surveys [4, 15], discovered variants explain only a small proportion of the observed familial aggregation [2, 16], thus posing a famous problem of missing disease heritability [17]. While there are a few proposed solutions to the encountered challenge [5], an urgent contemporary question that still needs to be solved is regarding the architecture of complex human traits. While, "common variant" hypothesis has come under a lot of criticism lately [1, 17], it is now necessary to devise experimental and computational methods to determine which one of the proposed disease architectures describes the reality in order to help develop future clinical medicine applications of bioinformatics technologies [1, 3, 17].

The most common type of DNA change is known as the single-nucleotide polymorphism (SNP), which arises when a single base (A, T, C, or G) is replaced by another one at a specific DNA position. Some SNPs can directly lead to disease formation; others increase the chance of disease statistically [18]. Analysis of SNP data is complicated because of a large number of possible interaction combinations as well as by the presence of correlation with the nearby SNPs.

Beyond Single-Locus Analysis

Despite striking success in the twentieth century in pinpointing genes responsible for Mendelian diseases, genetic origins of common complex diseases are, in fact, non-Mendelian in nature [9, 19]. Particularly, gene–gene interactions are involved in many complex biological processes like metabolism, signal transduction and gene regulations; thus, genetic variants in multiple loci may contribute to the disease formation together [20, 21]. For example, breast cancer and type 2 diabetes have been linked to multi-SNP interactions [21–23]. While most current bioinformatics approaches focus on detecting single-SNP associations, advanced statistical methods are necessary for multi-SNP association mapping because single-variant methods not only lose power when interactions exist but are, in fact, helpless in detecting rare mutations [24]. Also, the number of possible interactions is so vast that it is computationally unrealistic to search through all possible interactions in the genome for a large-scale case-control study [8, 25].

Additional challenge for disease origin discovery comes from the statistical correlation between nearby variants known as linkage disequilibrium or LD [25, 26]. LD patterns have many important applications in genetics and biology [27] and arise due to shared ancestry for contemporary chromosomes [13]. Due to LD patterns, it is likely that there will be a lot of redundant positive signals in dense studies [24]. Later on we address in detail how Bayesian strategies can address the burning problems in genetics while dealing with epistasis and linkage disequilibrium.

Modern Bioinformatics Approaches

Currently, most of the approaches to disease association mapping employ the standard "frequentist" attitude to the evaluation of significance [2]. Particularly, such algorithms use hypothesis testing procedures to deal with one variant at a time [24]. However, failures of such "frequentist" methods to account for the power of a study and the number of likely true positives [2] combined with the increased likelihood to report a multitude of redundant associations [24] sparked a wide interest in the Bayesian procedures. In this review, we survey the challenges facing statistical geneticists while analyzing the GWAS data and outline how recently emerged Bayesian methods can help with the process. In addition to outlining the main differences between various proposed approaches, we highlight limitations and advantages of each method and describe future prospects in the field and how Bayesian approaches can aid in answering outstanding questions in biomedicine.

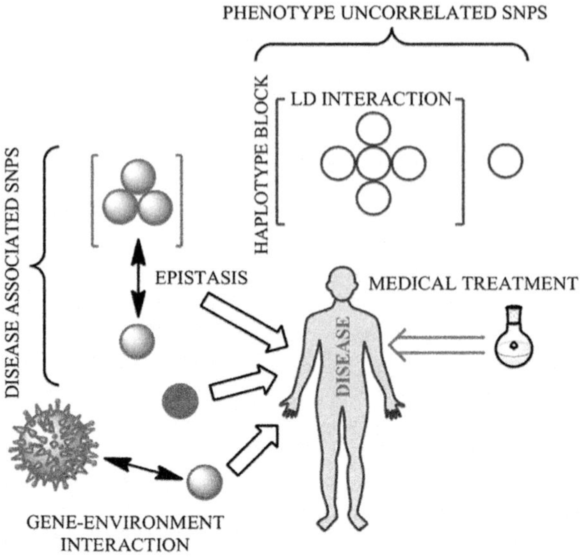

Fig. 1 Genetic interaction graph showing possible paths to disease formation. SNPs are represented as *circles* with *color* and *shading* indicating disease connection: "*red solid*" ones are marginally associated with the phenotype under consideration, "*red shaded*" ones are leading to disease formation through epistasis or are in linkage disequilibrium (LD) with such variants, and finally "*green circles*" are not associated with the phenotype. LD blocks are shown as *square brackets* and interactions are depicted as *double-headed arrows*

Bayesian Data Analysis Methods

In Fig. 1, we have shown multiple complicated interactions that have to be considered while developing statistical models for understanding of the multi-locus interactions resulting in the disease development. The ultimate goal is to be able to accurately understand all the shown connections in large-scale case-control studies while also comprehending the biological processes that lead to disease development. Thus, while statistical understanding is important, developing methods that can point in the direction of the appropriate biological processes taking place is the ultimate goal.

Overview of Bayesian Data Analysis

Statistical conclusions about an unknown parameter θ (or unobserved data x_{unobs}) in the Bayesian approach to parameter estimation are described utilizing probability statements, which are conditional on the observed data x: $p(\theta|x)$ and $p(x_{unobs}|x)$. Additionally, implicit conditioning is performed on the values of any covariates

[28]. The concept of conditioning on the observed data is what separates Bayesian statistics from other inference approaches which estimate unknown parameter over the distribution of the possible data values while conditioning on the true, yet unknown parameter values [28, 29].

At the heart of all the Bayesian approaches for detection of gene–gene interactions lies the concept of Bayesian inference and model selection. The goal is to determine the posterior distribution of all parameters in the problem (disease association, epistatic interactions, gene–environment interactions and others), given the common variants data for the case-control study while incorporating prior believes about parameter values. The conditional probability of all parameters (Params) given the observed data (Data) is given by the product of the likelihood function of the data and prior distribution on the parameters, as well as the normalization constant:

$$P(\text{Params}|\text{Data}) = \frac{P(\text{Data}|\text{Params})P(\text{Params})}{P(\text{Data})} \quad (1)$$

For most high-dimensional data sets encountered in large-scale studies, $P(\text{Data})$ cannot be explicitly calculated [9] and, therefore, $P(\text{Params}|\text{Data})$ can be evaluated analytically only up to the proportionality constant. However, advanced computational techniques (iterative sampling methods) can be used to determine posterior distribution of parameters [29, 30]. The main task is to make appropriate choices of statistical models to describe the likelihood expression and also to choose appropriate prior distributions on the values of parameters, $P(\text{Params})$.

Overview of Bayesian Variable Partition

Instead of testing each SNP set in a stepwise manner [31, 32], Bayesian approaches fit a single statistical model to all of the data simultaneously [9, 25, 33] allowing for increased robustness when compared to hypothesis testing methods [2, 24]. Another advantage of Bayesian approach to the problem is the ability to quantify all the uncertainties and information, and to incorporate previous knowledge about each specific SNP marker into the statistical model through priors [9, 29].

In the Bayesian model selection framework, we are interested in figuring out which of the set of models $\{M_i\}_{i=1}^{N}$ is the most likely one given the observed Data. The posterior probability for a particular model M_i given Data is described by:

$$P(M_i|\text{Data}) \propto P(\text{Data}|M_i)P(M_i) \quad (2)$$

Thus, through comparison of the posterior odds ratio for $P(M_i|\text{Data})$ and $P(M_j|\text{Data})$ it can be determined whether model M_i or M_j is more likely [29]. It is important to note that the normalization constant in Eq. 2 involves summation over all possible models: $P(\text{Data}) = \sum_{i=1}^{N} P(\text{Data}|M_i)P(M_i)$. For example, consider the case of a genome-wide study containing 1500 SNPs each of which can take one of the three possible states; thus, $N = 3^{1500} \approx 5 \times 10^{715}$ is the total number of feasible models to sum over. In such instances, it is necessary to use stochastic methods to sample from the posterior distribution. Now let us consider how this conceptual framework is applied in practice to the determination of multi-locus interactions in case-control studies.

Epistasis Analysis Methods

While statistical methods like BGTA [34], MARS [35], and CPM [36] are capable of detecting epistatic associations, the Bayesian epistasis association mapping (BEAM) algorithm [9] was the first practical approach capable of handling genome-wide case-control data sets. BEAM algorithm gives for each SNP marker posterior probabilities for disease association and epistatic interaction with other markers given the case-control genotype SNP data. Figure 2 shows the input file format necessary for application of the algorithm. The core of the Bayesian marker partition model used can be briefly summarized as follows.

BEAM can detect both interacting and noninteracting disease loci among a large number of variants. It is an application of Bayesian model selection procedure. Particularly, all the markers are split into three groups: (1) markers not associated with the disease, (2) marginally disease-associated variants, and (3) those with interaction associated disease effect. Thus, using the priors on the marker memberships and Markov Chain Monte Carlo (MCMC) methods, posterior probabilities for group memberships are determined. Specifically, by interrogating each SNP marker conditionally on the current status of others via MCMC method, the algorithm produces posterior probabilities [9]. Particularly, the genotype counts

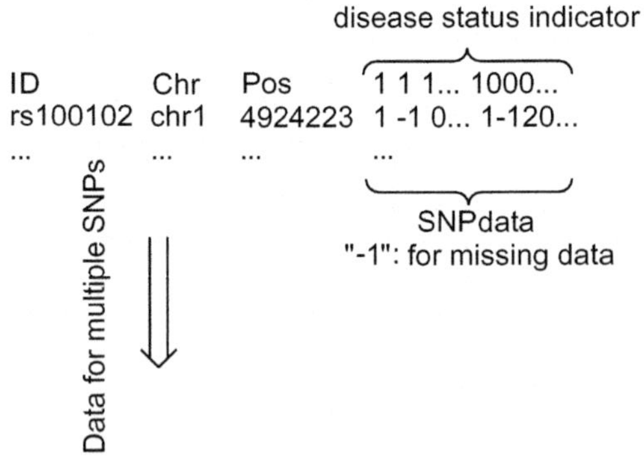

Fig. 2 Input data format for BEAM software [9] which uses MCMC to analyze case-control genetic studies. Label "1" denotes patients while "0" denotes controls. Note that it is not a requirement to provide the SNP ID and location

are modeled by the multinomial distribution with frequency parameters $\theta = \{\theta_1, \theta_2, \theta_3\}$, $\sum_{i=1}^{3} \theta_i = 1$ described by the Dirichlet prior:

$$P(\theta|\alpha) \propto \prod_{i=1}^{3} \theta_i^{\alpha_i - 1} \tag{3}$$

In order to determine the posterior probability of each marker's group membership (represented by I), the Metropolis–Hastings (MH) algorithm [30] is used to sample from $P(I|D,H)$ as given in Eq. 3:

$$P(I|D,H) \propto P(D_1|I)P(D_2|I)P(D_0,H|I)P(I), \tag{4}$$

where D is the patient data set (with disease), H is the control data set (healthy), and then D_0, D_1, and D_2 are correspondingly partitions of the patient data set into the three categories described above. The assumption is that case genotypes at the disease-associated markers will have different distributions when compared to control genotypes. Furthermore, the likelihood model assumes independence among markers in control group.

While BEAM algorithm was one of the first few to be able to handle GWAS data, it suffered from an assumption that SNPs dependence structure could be described by the Markov chain [9, 25]. In fact, SNP markers are highly correlated within haplotype blocks which are separated by recombination events [13, 37]. Therefore, despite its success, BEAM model is unable to capture the block-like human genome structure.

Incorporating Block-Type Genome Structure

Given that nearby SNPs are strongly correlated due to linkage disequilibrium, a new Bayesian model [25] that infers diplotype blocks and chooses SNP markers within blocks that are disease-associated becomes much more powerful when compared to other similar approaches. Here, we review the statistical Bayesian model for the LD-block structure determination [25, 26]. The main assumption is that diplotypes of individuals come from a multinomial distribution with frequency parameters described by the Dirichlet prior and that genotype combinations of SNPs in different blocks are mutually independent. The compact expression for the marginal probability of the data for a specific block is given by:

$$P\left(D_{[s,b)}|[s,b) = block\right) = \left(\prod_{i=1}^{3^{b-s}} \frac{\Gamma(n_i + a_i)}{\Gamma(a_i)}\right) \frac{\Gamma(\sum a_i)}{\Gamma(\sum(n_i + a_i))}, \tag{5}$$

where a block of SNPs considered consists of the SNPs (s, ..., $b-1$); Γ is the gamma function, $\vec{a}$ is the vector of Dirichlet parameters and *ni* refers to the number of counts for a specific diplotype. For joint inference of diplotype blocks and disease association status, we use the joint statistical model for the observed genotype data in cases and controls, the marker membership and block partition variable:

$$P(H, D, B, I) = P(H, D|B, I)P(B)P(I) \tag{6}$$

Finally, in order to determine the posteriors $P(B|D, H)$ and $P(I|D, H)$ the model uses a combination of MH algorithm and Gibbs sampler [25].

Detailed Interaction Partition Structure Determination

While successful in inferring epistatic interactions in large-scale case-control studies, both BEAM and its newer version BEAM2 had a disadvantage of using saturated models which limited the ability of the algorithms to accurately determine the epistatic interactions structure. Recent studies showed [4, 33, 38] that such interaction details arising due to encoding of the complicated regulatory mechanisms might play an important role in the disease formation. In order to carefully explore the etiopathogenesis and genetic mechanisms of diseases, a novel algorithm named Recursive Bayesian Partition (RBP) was proposed [33]. The RBP approach employs a Bayesian model to discover independence groups among interacting markers: first, it recursively infers all the marginally independent interaction groups, and then determines the conditional independence within each group using a chain dependence model. RBP therefore successfully recursively determines dependence structure among interacting variants in GWAS. Figure 3 shows an example of the possible outcomes of the RBP algorithm applied to GWAS data when determining the epistatic interactions independence structure.

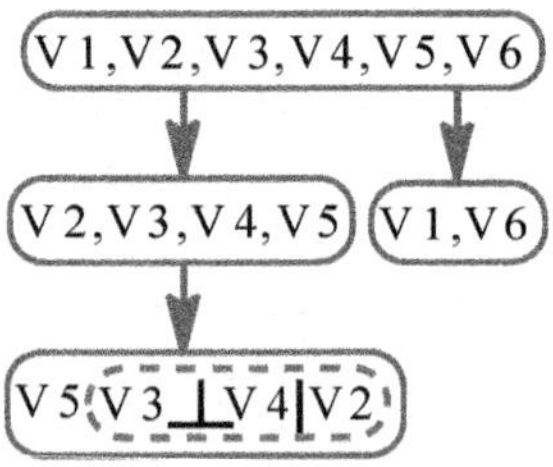

Fig. 3 Inference of the detailed dependence structure using Recursive Bayesian Partition (RBP) method. The individual independence groups among five variants are pointed out using separate *solid blocks*. A group with conditional independence is denoted using a *dotted shape*

Bayesian Graph Models and Networks

In order to improve disease mapping sensitivity and specificity, BEAM3 algorithm [24] uses a graph model to allow for flexible interaction structures for multi-SNP associations. Through the use of Bayesian networks, BEAM3 detects flexible interaction structures instead of using saturated models (like BEAM and BEAM2), therefore, highly reducing the interaction model complexity. Moreover, because only the disease association graphs are constructed, BEAM3 provides for higher computational efficiently in the whole-genome association settings [24].

In detail, BEAM3 allows for higher order couplings via saturated interactions within cliques (nonoverlapping partition of SNPs) and pairwise interactions between them. It can be shown [24] that the joint probability of all SNPs *X*, parameters, including disease graph and association status (*G*, *I*), and disease status indicator (*Y*) is given by:

$$P(X, Y, G, I) \propto P_A(X_1|Y, G)P(Y)P(G|I)P(I)/P_0(X_1), \tag{7}$$

where $G = (C, \Delta)$ is an undirected disease graph constructed on disease-associated SNPs (X_1) and including partition of SNPs into cliques (*C*) and interaction between cliques (Δ); probability function of X_1 set under the phenotype association hypothesis is described by P_A. Therefore, as can be seen from Eq. 7, only a few disease-associated SNPs are modeled (in set X_1), and hence a significant portion of computational time is saved by avoiding explicit modeling of complicated dependence structures of all SNPs which could be millions [19, 24, 39]. Additionally, through the choice of a proper baseline probability function $P_0(X_1)$, the model automatically accounts for the complex LD effects among dense SNPs employing graphs. Thus, a significant number of repetitive false interactions are avoided reducing computational burden [24]. Specifically, summing over all G' graphs, the expression for the baseline model becomes:

$$P_0(X_1) = \sum_{G'} P_0(X_1|G')P(G') \tag{8}$$

An alternative approach toward learning disease inducing gene–gene interactions is using binary classification trees. Bayesian methodology has been recently applied [21] to identification of multi-locus interactions in the large-scale data sets using a Bayesian classification tree model. Specifically, this kind of machine learning approach produces tree structure models, where each nonterminal node determines the splitting rule based upon the predictor variables like SNPs, and edges between nodes correspond to different possible values for the variable in the top parent node. A path along such a tree till the terminal node represents a specific combination of predictor variables along the path, therefore, automatically accommodating for epistasis [8, 21].

There are various ways for searching through the feasible tree space in such recursive partitioning approaches including greedy algorithms [40], random forests approach [8, 41], and MCMC [42, 43]. Bayesian variable partition and Bayesian classification trees are conceptually similar in that prior is assigned to all the tree models with the purpose of controlling the tree size [21]. One main advantage of this approach is in a possible enhancement of finding probability for multi-locus interactions with weak marginal effects due to ensuring the variable splitting through the prior specification. Moreover, due to the adaptivity of the MCMC algorithm, such Bayesian tree models detect higher order interactions by performing thorough searches near trees with the interacting variables determined in previous iterations [21]. It is important to point out that classification tree approaches do not test for epistatic interactions directly [8].

Clinical Applications of Bayesian Methodology

Even though practical Bayesian approaches for whole-genome multi-locus interactions analysis have emerged relatively recently, such methods have already helped to make important advances in determination of disease etiology. Table 1 succinctly summaries and compares all the statistical methods described above as

Table 1 Comparison of modern Bayesian approaches for whole-genome association analysis with possible clinical applications

SNP analysis method	Brief description	Interactions model	Detected loci/Epistatis
Bayesian Epistasis Association Mapping (BEAM) [9]	Epistasis detection	Saturated	More powerful than previous approaches
Bayesian Epistasis Association Mapping 2 (BEAM2) [25]	Epistasis/LD-block detection	Saturated	Many previous loci + new two-way associations
Recursive Bayesian Partition (RBP) [33]	Detailed independence structure of epistasis	Marginal and conditional independence groups	Confirmed previously known saturated interactions
BEAM3 [24]	Bayesian graph model for epistasis/LD[a]	Flexible graph structure	All previous IBD[b] loci + 2 new + 2 interchr.[c] interactions
Bayesian Classification Tree [21]	Classification tree model/recursive partitioning	Classification tree	Possible epistasis identified
BEAM + BEAM2 [44]	Interchromosomal epistatic interactions study	Saturated	319 high-order interactions found

As can be seen from the table, studies applying Bayesian methodology have already identified potential multi-locus interactions in high-dimensional datasets
[a]Linkage-disequilibrium; [b]Inflammatory bowel disease; [c]Interchromosome

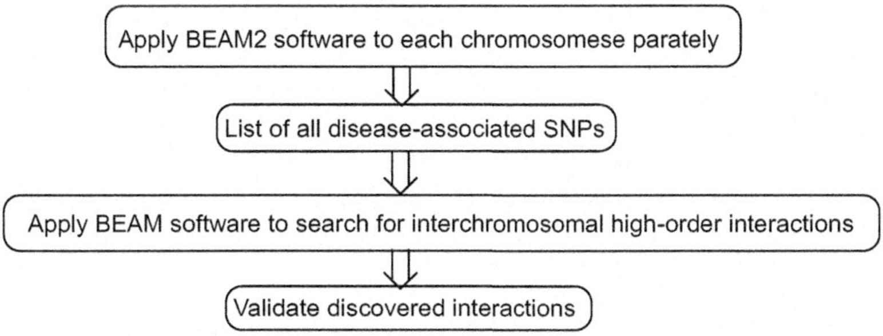

Fig. 4 A schematic diagram of the Bayesian analysis strategy combining multiple software applications [44]. In order to account for linkage disequilibrium, BEAM2 algorithm [25] was used to discover chromosome-wise interactions. However, a more efficient BEAM algorithm [9] was used on the determined SNPs across all chromosomes

well as their success in determination of the previously known disease loci and, more importantly, in the discovery of new multi-locus interactions responsible for complex diseases. We specifically note what interaction model each method utilizes. For example, Bayesian analysis strategy combining BEAM and BEAM2 software [44] allowed for the discovery of 319 high-order interactions across the genome that can potentially explain the missing genetic component of the rheumatoid arthritis susceptibility. Moreover, their findings indicate that nervous system, in addition to autoimmune one, potentially performs a crucial role in the disease development. Figure 4 shows a schematic diagram of the combined Bayesian strategy used for the analysis. This is an example of the statistical study in which disease underlying biological processes can be extracted from determined statistical associations. For sure, many more studies will follow in the near future that apply Bayesian methods either to existing GWAS data or to new large-scale studies.

Conclusions and Future Prospects

Certain issues need to be considered when using Bayesian approaches described above. For example, the combination of genotyping errors, disease heterogeneities, and population substructures could have adverse effects on the statistical results of the methods [9]. Currently, the major problem in the field is that the determined disease-associated genetic variants explain only a small part of the disease heritability [3, 4]. However, it is conceivable that the usage of the software tools outlined above will help with the detailed understanding of the interactions involved. Additionally, recent development of Bayesian models should allow for the elucidation of the detailed etiopathogenesis of the disease formation and the underlying causal biology.

Improvements to the Bayesian approaches mentioned in this article can include incorporation of environmental factors and population structures as covariates in the statistical model [33, 45]. Another possible improvement is to impute untyped SNPs and missing genotypes [46]. Efficient incorporation of prior biological knowledge into the Bayesian model can increase the probability of making discoveries in association studies [47]. Finally, recent computational proposals attempt to apply Bayesian methodology specifically toward efficient identification of causal rare variants in GWAS [48, 49].

It is important to keep in mind that the clinical applications of the statistical methods will arise from the understanding of the relationship between determined mathematical couplings and their biochemical underpinnings. The biological interpretation of the determined single- and multi-variant effects is currently a crucial area of research in genetics [8]. Modern statistical approaches to the analysis of the SNP data from whole-genome association studies have potential to play an important role in the future of bioinformatics and genomics research. Specifically, such methods will contribute to novel understandings of disease pathogenesis and provide crucial information for drug discovery [50], thus leading to important clinical applications.

Acknowledgements Zhang was supported by the start-up funding and Sesseel Award from Yale University.

References

1. S.S. Hall, Revolution postponed. Sci. Am. **303**, 60–67 (2010)
2. M.I. McCarthy et al., Genome-wide association studies for complex traits: consensus, uncertainty and challenges. Nat. Rev. Genet. **9**, 356–369 (2008)
3. P. Donnelly, Progress and challenges in genome-wide association studies in humans. Nature **456**, 728–731 (2008)
4. WTCCC, Genome-wide association study of 14,000 cases of seven common diseases and 3,000 shared controls. Nature **447**, 661–678 (2007)
5. E.E. Eichler et al., Missing heritability and strategies for finding the underlying causes of complex disease. Nat. Rev. Genet. **11**, 446–450 (2010)
6. J.A. Todd et al., Robust associations of four new chromosome regions from genome-wide analyses of type 1 diabetes. Nat. Genet. **39**, 857–864 (2007)
7. J.N. Hirschhorn, M.J. Daly, Genome-wide association studies for common diseases and complex traits. Nat. Rev. Genet. **6**, 95–108 (2005)
8. H.J. Cordell, Detecting gene-gene interactions that underline human diseases. Nat. Genet. **10**, 392–404 (2009)
9. Y. Zhang, J.S. Liu, Bayesian inference of epistatic interactions in case-control studies. Nat. Genet. **39**, 1167–1173 (2007)
10. M.L. Metzker, Sequencing technologies—the next generation. Nat. Rev. Genet. **11**, 31–46 (2010)
11. D. Branton et al., The potential and challenges of nanopore sequencing. Nat. Biotechnol. **26**, 1146–1153 (2008)
12. A. Schaffer, Nanopore sequencing. Technol. Rev. (2012)

13. The International HapMap Consortium, A haplotype map of the human genome. Nature **437**, 1299–1320 (2005)
14. E. Svoboda, The DNA transistor. Sci. Am. **303**, 46 (2010)
15. A.D. Johnson, C.J. O'Donnell, An open access database of genome-wide association results. BMC Med. Genet. **10**, 6 (2009)
16. D. Altshuler, M. Daly, Guilt beyond a reasonable doubt. Nat. Genet. **39**, 813–815 (2007)
17. G. Gibson, Rare and common variants: twenty arguments. Nat. Rev. **13**, 135–145 (2012)
18. M. Carmichael, One hundred tests. Sci. Am. **303**, 50 (2010)
19. X. Jiang et al., Learning genetic epistasis using Bayesian network scoring criteria. BMC Bioinform. **12**, 89 (2011)
20. J.H. Moore, The ubiquitous nature of epistasis in determining susceptibility to common human diseases. Hum. Hered. **56**, 73–82 (2003)
21. M. Chen et al., Detecting epistatic SNPs associated with complex diseases via a Bayesian classification tree search method. Ann. Hum. Genet. **75**, 112–121 (2011)
22. M.D. Ritchie et al., Multifactor-dimensionality reduction reveals high-order interactions among estrogen-metabolism genes in sporadic breast cancer. Am. J. Hum. Genet. **69**, 138–147 (2001)
23. S. Wiltshire et al., Epistasis between type 2 diabetes susceptibility loci on chromosomes 1q21-25 and 10q23-26 in Northern Europeans. Ann. Hum. Genet. **70**, 726–737 (2006)
24. Y. Zhang, A novel graphical model for genome-wide multi-SNP association mapping. Genet. Epidemiol. **36**, 36–47 (2012)
25. Y. Zhang et al., Block-based Bayesian epistasis association mapping with application to WTCCC type 1 diabetes data. Ann. Appl. Stat. **5**, 2052–2077 (2011)
26. I. Kozyryev, J. Zhang, Bayesian determination of disease associated differences in haplotype blocks. Am. J. Bioinform. **1**, 20–29 (2012)
27. J.D. Wall, J.K. Pritchard, Haplotype blocks and linkage disequilibrium in the human genome. Nat. Rev. Genet. **4**, 587–597 (2003)
28. A. Gelman et al., *Bayesian Data Analysis*, 2nd edn. (2003)
29. J.A. Rice, *Mathematical Statistics and Data Analysis*, 3rd edn. (2006)
30. J.S. Liu, *Monte Carlo Strategies in Scientific Computing*, 1st edn. (2001)
31. J. Marchini et al., Genome-wide strategies for detecting multiple loci that influence complex diseases. Nat. Genet. **37**, 413–417 (2005)
32. Y. Liu et al., Genome-wide interaction-based association analysis identified multiple new susceptibility loci for common diseases. PLoS Genet. **7**, 3 (2011)
33. J. Zhang et al., A Bayesian method for disentangling dependent structure of epistatic interaction. Am. J. Biostat. **2**, 1–10 (2011)
34. T. Zheng et al., Backward genotype-trait association (BGTA)—based dissection of complex traits in case-control design. Hum. Hered. **62**, 196–212 (2006)
35. N.R. Cook et al., Tree and spline based association analysis of gene-gene interaction models for ischemic stroke. Stat. Med. **23**, 1439–1453 (2004)
36. M.R. Nelson et al., A combinatorial partitioning method to identify multilocus genotypic partitions that predict quantitative trait variation. Genome Res. **11**, 458–470 (2001)
37. D.E. Reich et al., Linkage disequilibrium in the human genome. Nature **411**, 199–204 (2001)
38. Y. Yang et al., Testing association with interactions by partitioning chi-squares. Ann. Human. Genet. **73**, 109–117 (2009)
39. Y. Zhang, J.S. Liu, Fast and accurate approximation to significance tests in genome-wide association studies. J. Am. Stat. Assoc. **106**, 846–857 (2011)
40. T. Hastie et al., *The Elements of Statistical Learning: Data Mining, Inference, and Prediction*, 5th edn. (2011)
41. L. Breiman, Random forests. Mach. Learn. **45**, 5–32 (2001)
42. H.A. Chipman et al., Bayesian CART model search. J. Am. Stat. Assoc. **93**, 935–948 (1998)
43. D.G.T. Denison et al., A Bayesian CART algorithm. Biometrika **85**, 363–377 (1998)
44. J. Zhang et al., High-order interactions in rheumatoid arthritis detected by Bayesian method using genome-wide association studies data. Am. Med. J. **3**, 56–66 (2012)

45. I. Lobach et al., Genotype-based association mapping of complex diseases: gene-environment interactions with multiple genetic markers and measurement errors in environmental exposures. Genet. Epidemiol. **34**, 792–802 (2010)
46. Y. Zhang, Bayesian epistasis association mapping via SNP imputation. Biostat **12**, 211–222 (2011)
47. M. Chen et al., Incorporating biological pathways via a Markov random field model in genome-wide association studies. PLoS Genet. **7**(4), e1001353 (2011)
48. F. Liang, M. Xiong, Bayesian detection of causal rare variants under posterior consistency. PLoS ONE **8**(7), e69633 (2013)
49. M.A. Quintana et al., Incorporating model uncertainty in detecting rare variants: the Bayesian Risk Index. Genet. Epidemiol. **35**, 638–649 (2011)
50. Y. Okada et al., Genetics of rheumatoid arthritis contributes to biology and drug discovery. Nature **506**, 376–381 (2013)

Imaging Genetics: Information Fusion and Association Techniques Between Biomedical Images and Genetic Factors

Dongdong Lin, Vince D. Calhoun and Yu-Ping Wang

Abstract The development of advanced medical imaging technologies and high-throughput genomic measurements has enhanced our understanding of their interplay as well as their relationship with human behavior. In this chapter, we review the recent work of fusing imaging and genetic data for the correlative and association analysis as well as the diverse statistical models in these studies from univariate to multivariate methods. We also discuss future directions and challenges in integrative analysis of imaging and genetic data and finally give an example of parallel independent component analysis (ICA) in an imaging genetic study of schizophrenia.

Background

Imaging genetics, as a new field to bridge imaging with genetics, aims to identify genetic factors that influence the intermediate quantitative measures from anatomical or functional images, and further the cognition and psychiatric disorders in humans. It provides a comprehensive understanding of neurobiological systems from genetic mutations to cellular processes, to the system level changes (e.g., brain

D. Lin · Y.-P. Wang (✉)
Biomedical Engineering Department, Tulane University, New Orleans, LA 70118, USA
e-mail: wyp@tulane.edu

D. Lin
e-mail: dlin5@tulane.edu

Y.-P. Wang
Center of Genomics and Bioinformatics, Tulane University, New Orleans, LA 70118, USA

D. Lin · V.D. Calhoun
The Mind Research Network & LBERI, Albuquerque, NM 87106, USA
e-mail: vcalhoun@unm.edu

V.D. Calhoun
Department of Electrical and Computer Engineering, University of New Mexico, Albuquerque, NM 87131, USA

D. Xu et al. (eds.), *Health Informatics Data Analysis*, Health Information Science,
DOI 10.1007/978-3-319-44981-4_7

structure, function and integrity), and eventually to human behavior. Many complex psychiatric disorders are heritable. The merge of imaging and genetics will facilitate the early diagnosis of psychiatric disorders, understanding of their pathophysiology and improvement of treatment in a personalized manner.

In recent imaging genetic studies, neural imaging measurements or endophenotypes are commonly used for genetic analysis [1, 2], i.e., brain structure and change, functional variation, connectivity and network. Compared to the diagnostic results from self-reported and questionnaire-based clinical assessments, the imaging endophenotypes are usually considered to be closer to the biology of genetic function; therefore, the penetrance of genetic variation at this level will be higher, which helps to boost the causal variants detection power [3, 4]. Some quantitative endophenotypes derived from brain imaging are reproducible and reliable with high heritability, and can accommodate highly heterogeneous symptoms from patients in the same group. Moreover, due to the high resolution of brain imaging (e.g., structural magnetic resonance imaging (MRI)), we can explore the genetic influence on specific regions of interest (ROI) across the entire brain, which refines the understanding of underlying neurobiological mechanism of psychiatric disorders.

Challenges

There are many challenges in integrative analysis of imaging and genetic data. For example, (1) the high dimensionality of datasets. Both imaging data (e.g., structural or functional MRI) and genetic data (e.g., single nucleoid polymorphism (SNP) or gene expression) have thousands of features which are usually much larger than the number of subjects. Such an overdetermined system imposes a challenge for conventional statistical methods; (2) multiple interactions of features. There are interactions among various subsets of these data sets such as SNPs within a linkage disequilibrium (LD) block, genes with co-expression, and brain regions with functional connectivity. There are also some nonlinear relationships such as epistasis of gene–gene interaction, SNP–SNP interaction and SNP diagnosis interaction. These interactions need to be carefully considered in the modeling in order to accurately explain the heritability of a given endophenotype and (3) other factors such as heterogeneity of data sets, environmental effects and rare causal variants [4, 5]. Therefore, a powerful statistical method is demanded. Currently, most imaging genetics studies are divided into four major categories: candidate gene-candidate phenotype analysis, candidate gene-whole brain analysis, candidate phenotype-whole genome analysis, and whole brain-whole genome analysis. We will review and summarize these methods in the following sections.

Current Techniques

Recent methodological developments in imaging genetics have evolved from candidate approaches to whole genome and whole brain methods, which we will focus on. The advantages and disadvantages of these methods are summarized in Table 1. For candidate approaches, they start from the candidate gene-candidate phenotype analysis. Candidate approaches are usually used to test a hypothesis regarding how the genetic variants influence the target imaging phenotype at specific brain regions [6]. It is costly and potential association may be missed due to incomplete prior knowledge. Candidate gene, whole brain analysis is to construct a statistical brain-wide parametric map by performing significant association test between each brain measure and candidate genetic variant. Such an approach is also based on high resolution of imaging for localization [7]. Candidate phenotype-whole genome analysis is a typical genome-wide association (GWA) study on quantitative imaging phenotype. The overall significance of each genetic variant effect is assessed by a genome-wide correction for multiple comparisons [8]. Usually Bonferroni-corrected p-values (e.g., often cited genome-wide significance level 5×10^{-8}) are used to select significant genetic variants due to the multiple testing. A large sample size is needed to improve the detection power and meta-analysis can also be performed to get more robust evidence [9].

The brain-wide, genome-wide association analysis is of great interest, however challenges for data analysis arise due to their high dimensionality and complexity. As shown in Fig. 1, there are mainly three types of methods proposed in current works: univariate-imaging and univariate-genetic association analysis (Fig. 1a), univariate-imaging and multivariate-genetic association analysis (Fig. 1b), and multivariate-imaging and multivariate-genetic association analysis (Fig. 1c).

A. Univariate-imaging univariate-genetic association analysis

This is an unbiased but exhausted method by performing independent test on each pair of imaging measurements and genetic variants. Stein et al. proposed a voxel-wise genome-wide association study (vGWAS) to screen each pair of SNP and voxel in maps of regional brain volume calculated by tensor brain morphometry (TBM) [10, 11]. This extensive searching results in a total of more than 10^{10} statistic tests with 27 h running on 500 CPUs. FDR was used to correct for multiple comparisons across the image and detect the effective variants [12]. Several top SNPs were found to be interesting; however, numerous multiple comparisons decreased the detection power dramatically so that none significant association could be found. Such a univariate-imaging univariate-genetic method ignores the spatial correlation of voxels in imaging data and LD structure along the genome, which will also decrease the power due to weak effect of single SNP. Instead of voxel level scan, Shen et al. [13] explored the genetic influence on ROI-based brain measures by averaging the voxel-based morphometry (VBM) values within ROI as

Table 1 Advantages and disadvantages of methods in imaging genetic study

Model	Selected methods	Advantages	Disadvantages
Candidate imaging/genetic factor analysis			
Candidate approaches	Candidate gene-candidate phenotype analysis/Candidate gene-whole brain-wide analysis/Candidate phenotype-whole genome analysis	Test specific brain regions or genes of interest; well established statistical methods	Need strong prior knowledge of interested features; the view of biological mechanism is limited
Whole brain genome-wide analysis			
Univariate imaging-Univariate genetic	vGWAS	No pre-filtering needed; no prior hypothesis; well established statistical methods	Computational extensive; multiple testing; ignore inter-collection among features; usually need large sample size
Univariate imaging-multivariate genetic	Set based test: GSEA-SNP GSA-SNP Plink set based test	Incorporate the LD of SNPs; reduce the dimension and the number of multiple comparison; computational complexity is reduced	Need prior knowledge on grouping features; need pre-filtering of features; spatial correlation of imaging phenotype is not considered
	Regression based method: vGeneWAS PCReg Regularized regression with group lasso, sparse group LASSO penalties		
Multivariate imaging-multivariate genetic	Two block analysis: Parallel ICA, pICA with reference, three-way pICA Regularized CCA/PLS	Consider the inter-collections among both imaging and genetic features; reduce the feature dimension and multiple comparison	Involve some parameters for model selection; potential over-fitting issue; results may be hard to interpret; need prior knowledge on grouping features
	Multivariate sparse regression methods: Sparse reduced rank regression Sparse multitask regression		

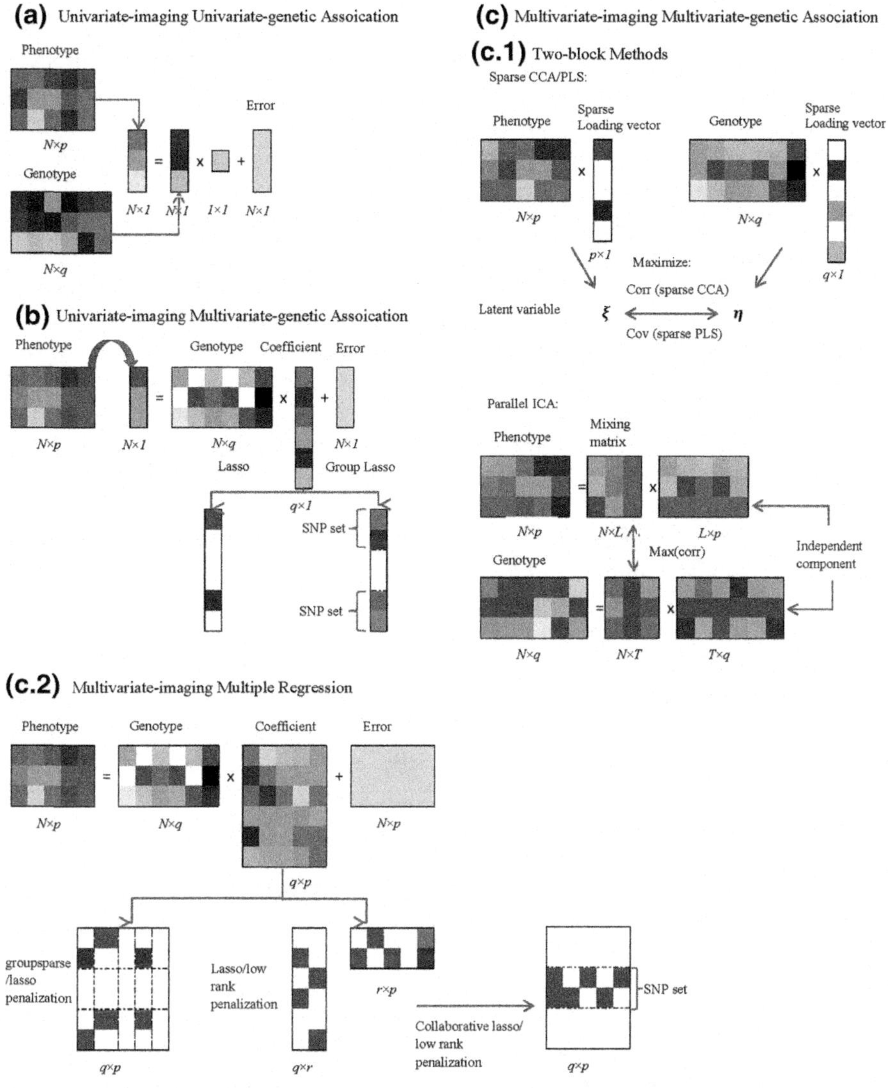

Fig. 1 Schematic illustration of brain-wide and genome-wide association analysis

traits for GWA scans. This ROI-based genome scan can reduce the effect of multiple comparisons and identify some significant associations, however, there may be only part of voxels within ROI having genetic factors and the simple average will induce noise into the imaging phenotype for detecting genetic biomarkers. Therefore, a more sophisticated multivariate method is needed to account for the intricate structure in both imaging and genetic features.

B. Univariate-imaging multivariate-genetic association analysis

Multivariate methods are commonly used to combine the effects of multiple SNPs as well as their interactions to model the joint effects on imaging phenotype. There are generally two ways to group SNPs: one is based on prior biological knowledge such as grouping SNPs from the same gene, pathway or network. The other is data-driven to group SNPs such as hierarchical clustering methods.

One way of constructing a SNP set-based test is inferring set-based test statistics or p-value from individual SNP test [14] such as a plink set-based test, gene set enrichment analysis (GSEA-SNP) and gene set analysis (GSA-SNP), as reviewed in [15]. Increase of sensitivity is expected for a SNP set-based test since it reduces the number of multiple comparisons and utilizes joint effect of SNPs with LD. However, the method may be sensitive to the way of grouping SNPs. In particular, if SNPs from the same group have weak LD and only some of them have causal effects, this set-based statistical analysis will not improve the performance.

An alternative way of testing the overall effect of SNP set is to construct a multiple linear regression (MLR) model. The challenges for MLR in imaging genetics are high dimensionality of genomic data and multi-collinearity between co-segregated SNPs within a LD block [16]. The linear system is underdetermined so that the estimates on the parameters are unreliable and sensitive to the collinearity among SNPs. A dimensional reduction technology such as principle component analysis (PCA) or sparse regression methods is needed. By PCA transformation, the SNPs will be projected to several orthogonal directions to form new regressors (principle components). These new regressors are ordered by their explained variance and can be applied for further association analysis, known as PC regression (PCReg). Chen et al. used PCA to reduce the dimension of almost 1 million SNPs before imputing them into a parallel ICA method [17]. Hibar et al. proposed voxel-wise gene-wide association study (vGeneWAS) to perform PCReg at each voxel [18]. They grouped SNPs based on gene membership and performed PCReg of each gene on each voxel to test the gene-wide association. The results show increased power and fewer tests than vGWAS but still none gene passing the multiple correction. Another voxel-wise GWAS method proposed by Ge et al. is a multi-locus model based on least squares kernel machines to associate the joint effect of multiple SNPs and their interactions on imaging traits [19].

Sparse regression methods have also gained increasing interests in recent years for approaching the large-scale data set. These include *l*-1 norm penalized regression (e.g., LASSO [20]) and elastic net [21], which could provide greater power by selecting a small number of genetic variants associated with imaging phenotype. Recently, in order to account for the group structure of genetic variants, some group-based penalization methods such as group LASSO (or structural LASSO) regression and sparse group LASSO regression are proposed. For example, Wang et al. [22] proposed a sparse multimodal multitask learning method with group LASSO penalty to identify associated SNPs for VBM, volumetric and cortical thickness, and memory scores. Silver et al. [23] applied a sparse group LASSO penalization regression method to identify genes and pathways related to high-density lipoprotein cholesterol. The results show better performance of the model than LASSO regression and a number of candidate gene/pathways were identified.

C. Multivariate-imaging and multivariate-genetic association analysis

As an extension of voxel-wide genetic association study, it is natural to consider the feature correlations or interactions between imaging and genetic data sets. One widely used approach is parallel ICA (pICA) [24]. pICA starts with PCA on both imaging measure and the SNPs. Then pICA is applied to both modalities to explore independent components from each modality respectively and maximize the correlation of the independent components between two modalities simultaneously. This is more powerful than univariate methods since the number of tests is reduced dramatically. Meda et al. [25] applied pICA to a whole brain genome-wide analysis. Liu applied pICA to identify those SNP components significantly associated with fMRI networks in schizophrenia and demonstrated that the selected components were discriminative. Extensions of pICA include pICA with reference [26] and 3-way pICA [27].

Other widely used methods include canonical correlation analysis (CCA) and partial least squares (PLS) with sparse penalizations to handle the high dimensionality and collinearity of data. Both methods assume that imaging and genetic data are linked by two latent variables (a linear combination of voxels and SNPs, respectively) and aim to estimate these latent variables by maximizing the correlation (CCA) or covariance (PLS) between the latent variables. Le Floch et al. [28] compared the performance of different methods such as univariate approach, sparse PLS, regularized kernel CCA and their combinations with PCA and pre-filters. The results show the best performance of filtering plus sparse PLS. Lin et al. proposed a group sparse CCA model for incorporating group structure into both imaging and genetic features. The performance of group sparse CCA is shown to be better than the existing sparse CCA methods. Two pairs of latent variables from imaging features and genetic variants in schizophrenia with significant correlation were identified [29].

Another promising approach is multivariate multiple regression, i.e., regressing the entire genetic variants on whole brain imaging measures with penalizations imposed on the coefficient matrix. Vounou et al. proposed a sparse reduced rank regression method (sRRR) to impose a low rank penalty on coefficient matrix and decompose it into two full-rank matrices which are also constrained to be sparse (l-1 norm) [30]. SRRR was tested to have higher sensitivity than univariate method and then was further applied to a whole brain-whole genome data set to identify those genetic variants associated with some AD-related imaging biomarkers [31]. Pathway sparse reduced rank, refined from sRRR, was developed to consider the joint effects of SNPs within the same pathway [32]. Lin et al. recently proposed a collaborative sparse reduce rank regression (c-sRRR) to incorporate more completed protein–protein interaction information into the grouping of SNPs and the method can perform bi-level selection on both SNP- and module-levels [33]. Several top genetic modules were identified to be associated with functional networks including postcentral and precentral gyri. In addition, sparse multimodal multitask regression [22] and group sparse multitask regression [34] proposed by Wang et al. can also be used to account for the group structure within predictors to improve the prediction performance.

Multivariate imaging-multivariate genetic methods have been shown to improve the detection power since it can search the whole brain and genome information, and account for the group structure among features in both modalities. Dimensional reduction is usually suggested to alleviate computational demand.

Conclusion

To summarize, imaging genetics is a new but promising field in exploring genetic effects on neurobiology and etiology of brain structure and function, and thus the human behavior and psychiatric disorders. A number of heritable imaging endophenotypes can improve our understandings of genetic influence on different brain patterns. Statistical methods in imaging genetics evolve from candidate approaches to whole brain-whole genome analysis, which enriches our discovery but also imposes challenges to develop powerful and efficient methods.

For the future imaging genetics studies, more types of genetic and epigenetic data (i.e., copy number variations and DNA methylation) are expected to integrate with brain imaging to provide more information in order to explain human behavior and their underlying biological mechanism. In addition, epistasis effect in genomic data and the interaction with environmental factors can be incorporated into the model to increase the detection power.

Example: Using Parallel ICA to Analyze fMRI and SNP Data Sets in Schizophrenia

The parallel ICA toolbox is developed based on MATLAB, which is freely available at http://mialab.mrn.org/software/fit/. FMRI and SNP data sets are also provided online as example datasets. There are 43 schizophrenia patients and 20 healthy controls under the folder named 'SZ' and 'Healthy' respectively. We will describe step-by-step approaches on how to perform pICA on these datasets.

Step A. Parameter setup

Open the interface shown in Fig. 2a, click the 'Setup Analysis' to select the output directory and then a window shown in Fig. 2b will pop up for setting up the parameters:

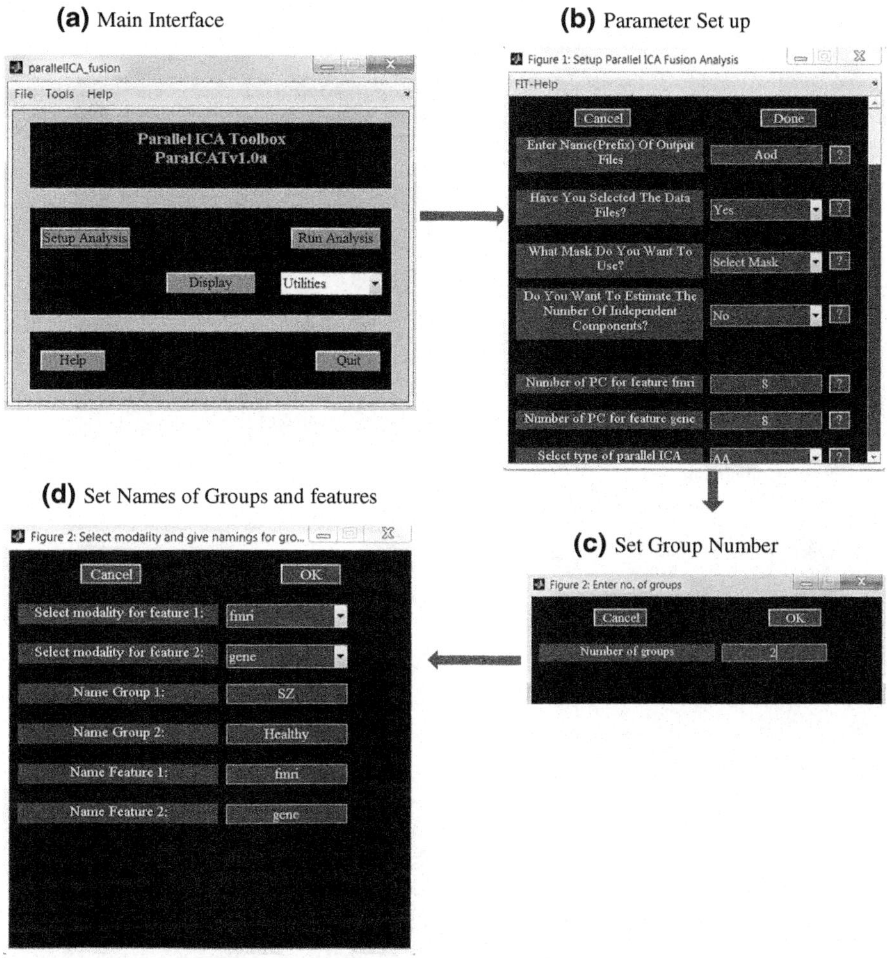

Fig. 2 Parameters set up for parallel ICA analysis

(1) Set 'Aod' as the prefix for all the output files.
(2) Set '2' for number of groups as shown in Fig. 2c and move to set the names of groups and features. Group names are 'SZ' and 'Healthy' and features are 'fmri' and 'gene' as shown in Fig. 2d. Then select the data files for each modality by using file pattern and directory selection. In the example data, the data files of two modalities from the same group are stored in the same folder and the file pattern (.img for fMRI data and .asc for SNP data) are the same in both groups. Finally, set the file directory for each group separately.
(3) Select the mask: 'myMask_t3.img' file for fMRI and indices (1:367) for SNP.

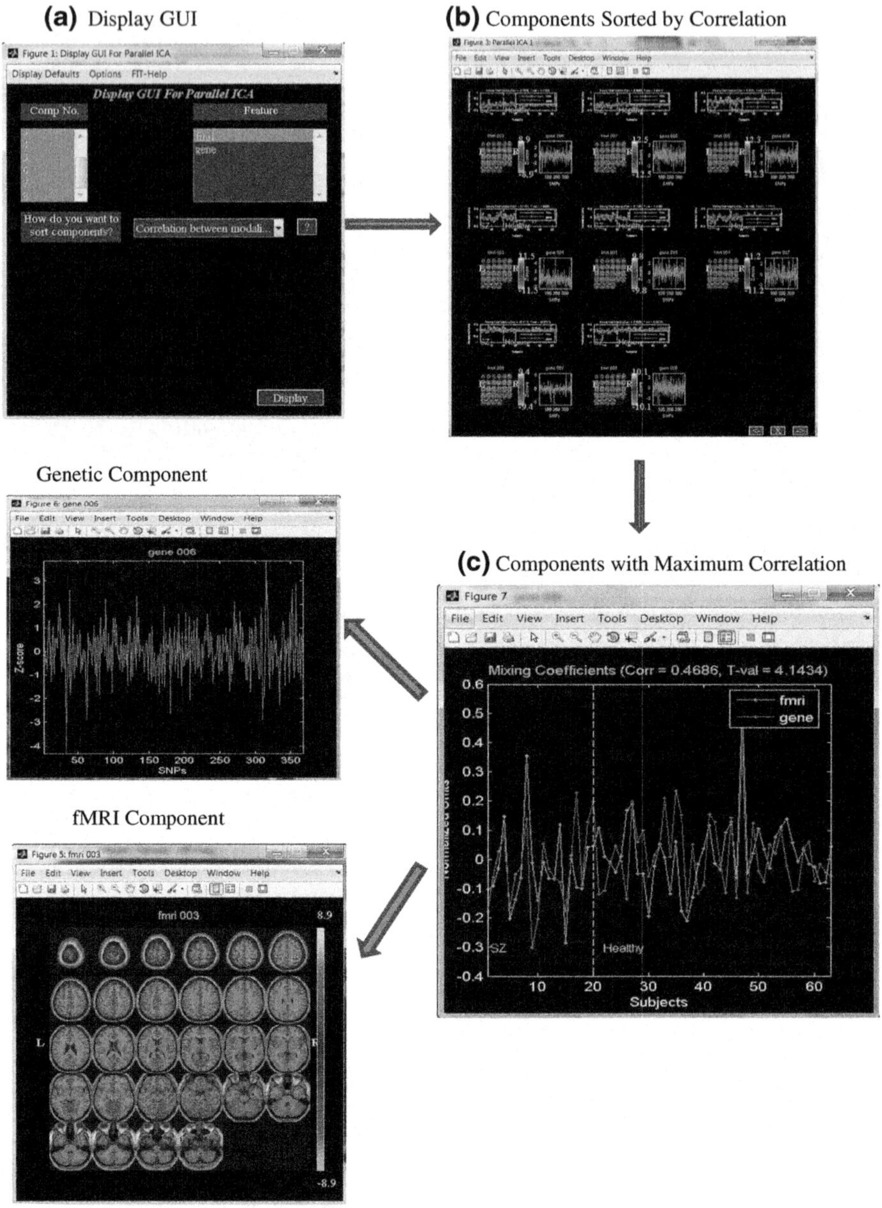

Fig. 3 Display of pICA results for fMRI and SNP analysis

(4) Set 8 as the number of principle components to be extracted from the modality fMRI and gene, respectively.
(5) Select Type of Parallel ICA as 'AA' for measuring correlation between mixing coefficients of two modalities.

(6) Select 'Reference' option for PCA with entering -1's for SZ group and 1's for healthy group.
(7) Select 'Average' as type of ICA and set 'Number of times ICA will run' as 10 to use averaged components across runs.

Step B. Run Analysis

Click 'Run Analysis' and select the parameter file ('Aod_para_ica_fusion.mat') to run pICA. The results are stored in 'Aod_para_ica_ica.mat' file.

Step C. Display

After the analysis is done, we can display the results by the component number or features.

We can sort 8 components by the measured correlation as shown in Fig. 3a. For each pair of components, fMRI and genetic components are shown as well as their mixing coefficients, correlation value and two sample t-test value on the pair of mixing coefficients. Figure 3b shows all the components extracted and Fig. 3c shows the pair of components with the largest correlation.

References

1. L. Shen et al., Genetic analysis of quantitative phenotypes in AD and MCI: imaging, cognition and biomarkers. Brain Imaging Behav. 1–25 (2013)
2. I.I. Gottesman, T.D. Gould, The endophenotype concept in psychiatry: etymology and strategic intentions. Am. J. Psychiatry **160**, 636–645 (2003)
3. A. Meyer-Lindenberg, D.R. Weinberger, Intermediate phenotypes and genetic mechanisms of psychiatric disorders. Nat. Rev. Neurosci. **7**, 818–827 (2006)
4. A. Meyer-Lindenberg, The future of fMRI and genetics research. Neuroimage **62**, 92–1286 (2012)
5. G. Northoff, Gene, brains, and environment—genetic neuroimaging of depression. Curr. Opin. Neurobiol. **23**, 133–142 (2013)
6. J.P. Andrawis et al., Effects of ApoE4 and maternal history of dementia on hippocampal atrophy. Neurobiol. Aging **33**, 856–866 (2012)
7. A.J. Ho et al., A commonly carried allele of the obesity-related FTO gene is associated with reduced brain volume in the healthy elderly. Proc. Natl. Acad. Sci. **107**, 8404–8409 (2010)
8. J.L. Stein et al., Genome-wide analysis reveals novel genes influencing temporal lobe structure with relevance to neurodegeneration in Alzheimer's disease. Neuroimage **51**, 542–554 (2010)
9. S.A. Melville et al., Multiple loci influencing hippocampal degeneration identified by genome scan. Ann. Neurol. **72**, 65–75 (2012)
10. J.L. Stein et al., Voxelwise genome-wide association study (vGWAS). Neuroimage **53**, 1160–1174 (2010)
11. A. Leow et al., Inverse consistent mapping in 3D deformable image registration: its construction and statistical properties, in *Information Processing in Medical Imaging* (2005), pp. 493–503
12. Y. Benjamini, Y. Hochberg, Controlling the false discovery rate: a practical and powerful approach to multiple testing. J. Roy. Stat. Soc. Ser. B (Methodol.), 289–300 (1995)

13. L. Shen et al., Whole genome association study of brain-wide imaging phenotypes for identifying quantitative trait loci in MCI and AD: a study of the ADNI cohort. Neuroimage **53**, 1051–1063 (2010)
14. J. Hoh et al., Trimming, weighting, and grouping SNPs in human case-control association studies. Genome Res. **11**, 2115–2119 (2001)
15. V.K. Ramanan et al., Pathway analysis of genomic data: concepts, methods, and prospects for future development. Trends Genet. **28**, 323–332 (2012)
16. K.A. Frazer et al., A second generation human haplotype map of over 3.1 million SNPs. Nature **449**, 851–861 (2007)
17. J. Chen et al., Multifaceted genomic risk for brain function in schizophrenia. Neuroimage **61**, 866–875 (2012)
18. D.P. Hibar et al., Voxelwise gene-wide association study (vGeneWAS): multivariate gene-based association testing in 731 elderly subjects. Neuroimage **56**, 1875–1891 (2011)
19. T. Ge et al., Increasing power for voxel-wise genome-wide association studies: the random field theory, least square kernel machines and fast permutation procedures. Neuroimage **63**, 858–873 (2012)
20. O. Kohannim et al., Discovery and replication of gene influences on brain structure using LASSO regression. Front. Neurosci. **6**, 115 (2012)
21. L. Shen et al., Identifying neuroimaging and proteomic biomarkers for MCI and AD via the elastic net, in *Multimodal Brain Image Analysis* (Springer, 2011), pp. 27–34
22. H. Wang et al., Identifying disease sensitive and quantitative trait-relevant biomarkers from multidimensional heterogeneous imaging genetics data via sparse multimodal multitask learning. Bioinformatics **28**, i127–i136 (2012)
23. M. Silver et al., Pathways-driven sparse regression identifies pathways and genes associated with high-density lipoprotein cholesterol in two Asian cohorts. PLoS Genet. **9**, e1003939 (2013)
24. J. Liu et al., Combining fMRI and SNP data to investigate connections between brain function and genetics using parallel ICA. Hum. Brain Mapp. **30**, 241–255 (2009)
25. S.A. Meda et al., A large scale multivariate parallel ICA method reveals novel imaging–genetic relationships for Alzheimer's disease in the ADNI cohort. Neuroimage **60**, 1608–1621 (2012)
26. J. Chen et al., Guided exploration of genomic risk for gray matter abnormalities in schizophrenia using parallel independent component analysis with reference. Neuroimage **83**, 384–396 (2013)
27. V.M. Vergara et al., A three-way parallel ICA approach to analyze links among genetics, brain structure and brain function. Neuroimage (2014)
28. E. Le Floch et al., Significant correlation between a set of genetic polymorphisms and a functional brain network revealed by feature selection and sparse Partial Least Squares. Neuroimage **63**, 11–24 (2012)
29. D. Lin et al., Correspondence between fMRI and SNP data by group sparse canonical correlation analysis. Med. Image. Anal. (2013)
30. M. Vounou et al., Discovering genetic associations with high-dimensional neuroimaging phenotypes: a sparse reduced-rank regression approach. Neuroimage **53**, 59–1147 (2010)
31. M. Vounou et al., Sparse reduced-rank regression detects genetic associations with voxel-wise longitudinal phenotypes in Alzheimer's disease. Neuroimage **60**, 700–716 (2012)
32. M. Silver et al., Identification of gene pathways implicated in Alzheimer's disease using longitudinal imaging phenotypes with sparse regression. Neuroimage **63**, 94–1681 (2012)
33. D. Lin et al., in *Proceedings of the 2013 IEEE International Conference on Bioinformatics and Biomedicine (BIBM),* Biomedical Engineering Department, Tulane University, New Orleans, LA, 70118, USA (2013), pp. 9–16
34. H. Wang et al., From phenotype to genotype: an association study of longitudinal phenotypic markers to Alzheimer's disease relevant SNPs. Bioinformatics **28**, i619–i625 (2012)

Biomedical Imaging Informatics for Diagnostic Imaging Marker Selection

Sonal Kothari Phan, Ryan Hoffman and May D. Wang

Abstract With the advent of digital imaging, thousands of medical images are captured and stored for future reference. In addition to recording medical history of a patient, these images are a rich source of information about disease-related markers. To extract robust and informative imaging markers, we need to regulate image quality, extract image features, select useful features, and validate them. Research and development of these computational methods fall under the science of biomedical imaging informatics. In this chapter, we discuss challenges and techniques of biomedical imaging informatics in the context of imaging marker extraction.

Introduction

Biomedical imaging informatics is a field of science that provides the computational means for handling, analyzing, and exploring images and their associated data to achieve a medical goal, e.g., diagnostic, therapeutic, or prognostic applications [1, 2]. Biomedical images can be broadly categorized into two groups: (1) macroscopic organ images such as MRI, CT, and PET and (2) microscopic tissue images such as histopathology, immunohistochemistry, and multispectral images. For analysis of these different types of images, researchers have developed and validated different imaging informatics methods. Therefore, several reviews and book chapters discuss research developed for different biomedical image types, including fluorescent microscopy [3–6], organ [7–9], and histopathology [1, 10–14].

S.K. Phan · R. Hoffman · M.D. Wang (✉)
The Wallace H. Coulter Department of Biomedical Engineering, Georgia Institute of Technology and Emory University, 313 Ferst Dr, UA Whitaker Bldg., Suite 4106, Atlanta, GA 30332, USA
e-mail: maywang@bme.gatech.edu

M.D. Wang
School of Electrical and Computer Engineering, Winship Cancer Institute, Parker H. Petit Institute for Bioengineering and Bioscience, Georgia Institute of Technology and Emory University, Atlanta, GA, USA

D. Xu et al. (eds.), *Health Informatics Data Analysis*, Health Information Science,
DOI 10.1007/978-3-319-44981-4_8

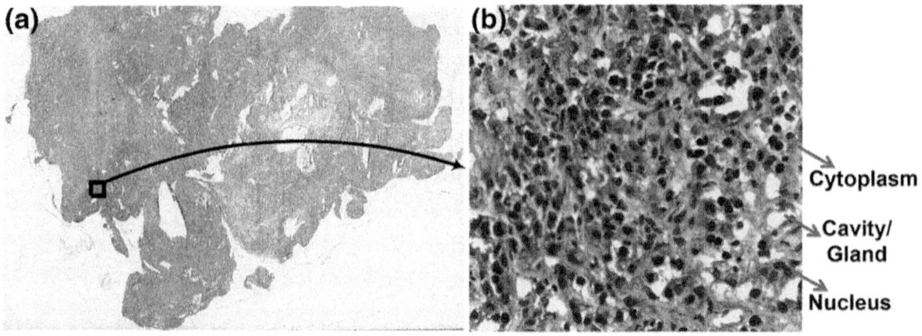

Fig. 1 A sample histopathological image stained with H&E stains. **a** WSI, and **b** 512x512-pixel rectangular section, where nuclei, cytoplasm, and glands appear *blue-purple, pink, and white*, respectively

Although informatics methods vary with the data characteristics, the basic components of the system are the same, including quality control, feature extraction, feature selection, and validation. In this book chapter, we will discuss specialized methods in these components for histopathological images.

Histopathology is the study of microscopic anatomical changes in diseased tissue samples. Tissue samples are usually obtained during surgery, biopsy, or autopsy and stained with one or more stains. Hematoxylin and Eosin (H&E) staining protocol is the most commonly used protocol for morphological analysis of tissue samples. H&E staining enhances four colors in histopathological images: blue-purple, white, pink and red (Fig. 1). These colors are associated with specific cellular structures. Basophilic structures—ribosome and nuclei—appear blue-purple; eosinophilic intra- and extracellular proteins in cytoplasmic regions appear bright pink; empty spaces—the lumen of glands—do not stain and appear white; and red blood cells appear intensely red. When studying histopathological images, pathologists study different types of image patterns such as shape of glands, density of nuclei, number of nucleoli, and morphology of cytoplasm (i.e., clear vs. granular cytoplasm).

Challenges

Image Artifacts

Errors in biopsy-slide preparation or microscope parameters may lead to anomalies, known as image artifacts, in histopathological images. Common image artifacts include tissue folds, blurred regions, pen marks, shadows, and chromatic aberrations [2, 14, 15]. These artifacts have unpredictable effects on image segmentation and feature extraction methods. Therefore, it is essential to either eliminate or correct these artifacts. Figure 2 illustrates pen marks and tissue folds in the whole-slide images (WSIs) from The Cancer Genome Atlas (TCGA) [16].

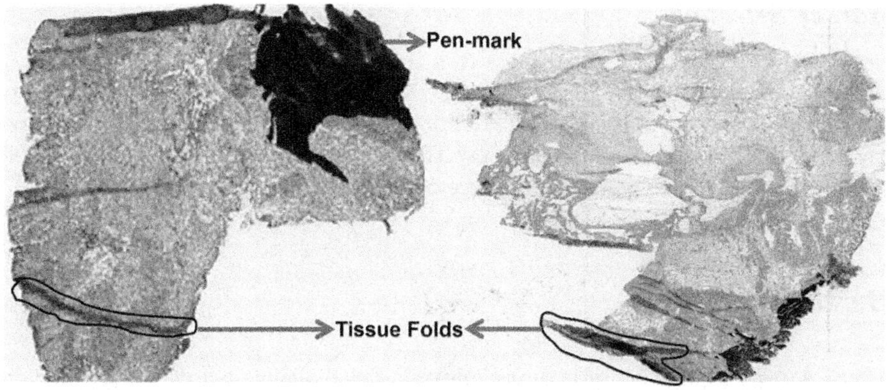

Fig. 2 Tissue fold and pen mark artifacts in whole-slide images provided by TCGA

Batch Effects

Differences in the setups used for slide preparation and acquisition systems between two batches of data may lead to differences in image appearance and properties between the two batches [2]. Histopathological images most commonly suffer from color and scale batch effects. Because of these batch effects, parameters and models optimized on one batch may not be optimal for other batches [17].

Object Detection

Image segmentation, the process of detecting objects in images, is an important step in pattern recognition applications. The performance of downstream feature extraction algorithms highly depends on the performance of image segmentation algorithms, which segment tissue regions, artifacts, stains, and nuclei. However, even after years of research, accurate image segmentation is a rather challenging task, which is further complicated by biological and technical variation in the data [18].

Semantic Gap

Image descriptors capture informative patterns that can predict disease using machine learning models. However, it is difficult to interpret these descriptors biologically because they are very different from the human interpretation of an image. This difference is often described as a semantic gap in the literature [9]. Because of this semantic gap, clinicians find it difficult to interpret and use the imaging markers.

Marker Selection

Feature extraction often results in thousands of image features due to the numerous feature extraction methods in the literature. Moreover, these image features often have multiple parameters and statistics. The challenge then is to select the most informative and robust features with limited available samples [19].

Marker Validation

After selecting a short list of imaging markers, the challenge is to validate them on larger datasets and establish them as biomarkers. Currently, most researchers only perform a cross-validation on a single batch of data with few samples. When applied to separately acquired data, most of these markers fail. It is essential to establish a strict marker validation guideline with rigorous cross-validation on a large dataset and blind validation on separately acquired dataset [19].

Current Techniques

Quality Control Methods for Addressing Image Artifacts and Batch Effects

Quality control methods extract useful tissue portions from images by eliminating or correcting image artifacts. Quality control is part of the data preprocessing methods, which often depend on the acquisition setups. Here, we discuss some commonly used quality control steps for histopathological images. If the images under study are WSIs, the first step of quality control is to locate tissue portions from the slide. Because tissue portions are colored and the remaining slide is white, they can be segmented using a saturation threshold for every pixel. The next step is to eliminate tissue folds caused by the layering of non-adherent tissue on the slide. Because of the tissue layering, pixels in fold regions have high saturation and low intensity. Tissue folds can be segmented by thresholding the difference value of saturation and intensity [20, 21]. Kothari et al. proposed an automated method that adaptively calculates optimal threshold for tissue-fold artifact detection [18]. Both of the above quality control steps should be performed on a low resolution image because these artifacts are evident on low resolution images and methods will run faster because of the smaller image size. The next step of quality control is to normalize color and brightness that may vary along the slide due to uneven slide illumination. Brightness can be normalized by subtracting a background function from the image [22] while the color of an image can be normalized to the color of a reference image [23–25].

Image Segmentation Methods for Object Detection

Image segmentation is an important step for object-level pattern recognition. Cellular structures in histopathological images can be easily distinguished based on stain colors. Therefore, structures can be segmented by classifying or clustering individual pixels using various color properties. Because of batch effects, stain colors may appear different from image to image. Therefore, for more accurate segmentation, researchers often use semi-automated methods, where the user selects colors or pixels of different structures [26, 27]. However, user interaction may introduce subjectivity in segmentation results. Some automatic segmentation methods overcome the color variation challenge by using more robust color properties and/or color normalization [28–30]. In addition to color properties, local texture properties, such as gradient, can further improve segmentation performance [31]. Since most objects are seldom made of a single pixel, segmentation performance can be further improved by considering pixel neighborhood properties using graph-cut [32], object-graph [33], and Markov models [34]. Characterizing object shape can further improve segmentation performance compared to basic pixel-based techniques. This is especially useful for accurate segmentation of nuclei, which are often characterized using elliptical shape models [35, 36]. Figure 3c illustrates a pseudo-colored segmentation mask and nuclear segmentation result for the histopathological image of kidney renal clear cell carcinoma tumor in Fig. 3a. In the segmentation mask, blue, pink, white, and red represent nuclear, cytoplasmic, no-stain/gland, and red blood cells regions, respectively.

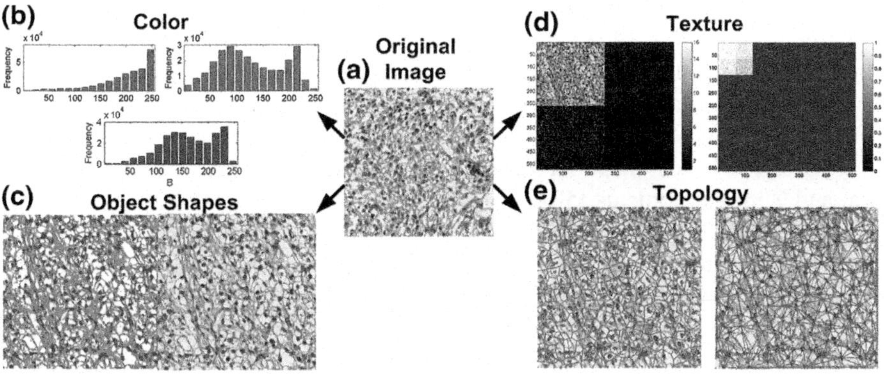

Fig. 3 Object segmentation and feature extraction in a portion of kidney renal clear cell carcinoma image. **a** Original image, **b** *Red, green, blue* color channel histograms, **c** Segmented stains and nuclear shapes, **d** Wavelet submatrices capturing information at different scales, **e** Delaunay triangulation and Voronoi diagrams using nuclear centers

Image Feature Extraction for Biologically Interpretable Markers

Color is an important aspect of pattern recognition in histopathological images, in which cellular structures appear in different colors. Color features extract color channels for every pixel in the image or for pixels in a region of interest. Although most images are represented in RGB color space (with red, green, and blue channels) [37, 38], they are often converted into different color spaces including HSV, LUV, and LAB [39–43]. Informative color features include color spread, prominence, and co-occurrence, which are captured using statistics and frequencies of color histograms (Fig. 3b).

Local variations in a flat plane or solid color of an image are characterized by texture properties such as image sharpness, contrast, changes in intensity, and discontinuities or edges. Different texture patterns emerge in cellular structures with disease progression; e.g., nuclear texture is informative for grading cancer [44, 45]. Texture features are usually extracted using grayscale images but recent work illustrated the utility of color texture, which was extracted from a color quantized image rather than grayscale quantized images [32]. Texture feature extraction methods include gray-level intensity profiles, Haralick Gray-level Co-occurrence Matrix (GLCM) features [27, 46, 47], properties of wavelet sub-matrices (Fig. 3d) [47, 48], properties of Gabor filter responses [46, 47], and Fractal dimensions [47]. Using a combination of these measures and parameters, informative image patterns can be captured for various image processing applications.

Shape cognition is an important aspect of human pattern recognition in images [49]. Pathologists routinely study nuclear and gland shapes for making their decisions. After object segmentation, shape-based features can be easily extracted using object contours and/or object regions (Fig. 3c) [49]. Contour-based features can either be extracted directly from shape boundaries or from parametric shape models such as Fourier shape descriptors and elliptical models. Contour-based features include model parameters, perimeter, boundary fractal dimension, and bending energy. Region-based shape features include solidity, Euler number, convex hull, area, and Zernike moments [50].

As a disease evolves, the spatial distribution of cellular structures often changes, e.g., when a normal tissue is invaded by tumor cells, nuclei multiply rapidly and nuclear density increases; after prolonged growth, tumor cells die in a local region leading to necrosis and decrease in nuclear density. These spatial patterns can be captured by topological or architectural features. To capture topology, spatial graphs such as Delaunay triangulations (Fig. 3e), Voronoi diagrams (Fig. 3e), minimum spanning trees, Gabriel graphs, and Ulam trees are often developed using centers of cellular objects as nodes [51, 52]. Thereafter, topological features are extracted from the graphs including edge length, connectedness, and compactness [53–55]. Topology can also be characterized using average distance between objects and number of objects within a given neighborhood [53].

Feature Selection for Imaging Marker Selection

It is important to identify the most informative markers in the feature set. One benefit of identifying the most informative markers or features is that it allows the dimensionality of the feature space to be reduced, simplifying models and improving overall performance. In addition, identifying the most informative features can provide insight into the system being modeled and inform future designs [56].

Feature selection algorithms can be subdivided into three classes: filters, wrappers, and embedded methods [19]. Filter methods can use a single statistical test or property for each feature and establish a threshold for rejection, or may consider multiple features together. Examples of filtering methods include t-tests, ANOVA, chi-square tests [57], minimum redundancy maximum relevance (mRMR) selection [58], and relief-F [59]. Filtering methods are computationally inexpensive as they do not require the training of a classifier for each feature or feature set, however this may also be a weakness in that the features are being selected independent of the specific classifier under consideration. Wrapper methods address this deficiency by generating sets of features and testing their performance directly using a classifier. Wrapper methods are typically applied in conjunction with search algorithms, such as sequential backward estimation, randomized hill climbing [60], genetic algorithms [61], or simulated annealing [62]. Wrapper methods suffer from a risk of over-fitting and are computationally expensive. Embedded methods seek to improve performance by using the classifier to identify the most important features, such as examining the weight vector of a SVM [63].

Classification for In-Silico Marker Validation

After feature selection, features often undergo in-silico validation using classification models. Classifiers are trained using a pathologist's annotations or other ground truth knowledge, combined with feature vectors whose true classes are known. Common classifiers used in biomedical applications are k-nearest neighbors (k-NN), support vector machines (SVM), Bayesian methods, neural networks, and decision trees [64]. It is difficult to predict which classifier will perform best for a specific application. As such, it is common to implement several of these algorithms and compare their accuracy [37, 47]. It is also possible to use boosting algorithms [46] or ensemble methods to combine multiple weaker classifiers and improve overall decision accuracy [42]. Cross-validation techniques should always be used when evaluating classifiers, so as to avoid biases from testing a classifier against its training data set [65].

Case Study: Imaging Marker Selection for Tumor Versus Nontumor Classification

A typical marker selection problem is differentiating between tumor and nontumor regions of histopathology images. In this case, an SVM classifier is trained to classify 512 × 512 segments of ovarian serous adenocarcinoma WSIs from the TCGA repository as tumor/nontumor using the following steps:

(1) Quality control algorithms are applied to eliminate from consideration any tiles that contain no tissue or image artifacts. First, any tile which contains more than 20% empty space or pen markings is eliminated. Next, tissue folds were identified and any tile found to contain 10% or more tissue fold by area was excluded from any further analysis.
(2) Features are extracted from artifact-free tissue tiles. In all, 461 features were extracted for each WSI tile including nine feature subsets as listed in Table 1 [66]. Color and global texture capture global image properties while other subsets capture object-specific biologically interpretable image properties.
(3) The next step is feature selection and validation. The mRMR method with two optimization functions—mutual information quotient and mutual information difference—and SVM classifier (LIBSVM) with linear kernel were used for feature selection and classification, respectively [67]. Feature vector length was considered in increments of 5, from 5 to 50 features. The following SVM cost weights were considered: 2^{-5}, 2^{-4},…, and 2^{10}. The internal loop of a 3-fold, 10 iteration nested cross-validation technique was used to select optimal feature size and SVM cost function weight, where preference was given to the lowest

Table 1 Contribution of different feature subsets in final feature list

Feature Type	Proportional representation in full feature set	Proportional representation in informative feature list	p-value (Fisher's exact test)
Color	0.15835141	31.3547619	9.30E-06
Global texture	0.299349241	16.8521164	0.9999008
Eosinophilic-object shape	0.110629067	5.232804233	0.994548
Eosinophilic-region texture	0.039045553	3.147883598	0.7881413
No-stain-object shape	0.110629067	0.207407407	1
Basophilic object shape	0.110629067	16.22063492	0.0587052
Basophilic-region texture	0.039045553	6.429100529	0.1735985
Nuclear shape	0.056399132	0	1
Nuclear topology	0.075921909	20.55529101	1.92E-06

feature size and the smallest cost [66]. Based on the model selected, the optimal feature set was selected using train set in external loop. After all 30 (3 fold, 10 iteration) external loops were completed, 30 optimal feature lists were generated. In large-scale experiments, this method has been found to have SVM classification accuracy of 95% or more [66].

(4) Optimal feature lists generated in the previous step had varying feature sizes. Therefore, percent contribution each subset was calculated for each external loop and averaged. Thereafter, a significance value was calculated for each subset using Fisher's exact test [66]. Table 1 shows the proportion of the informative feature list from each feature class.

Both color features and nuclear topology features were found to be significantly ($p < 0.001$) overrepresented in the informative feature sets for tumor/nontumor detection, meaning that those feature types had higher predictive power. On the other hand, neither nuclear shape nor no-stain object shape features appeared in any informative feature lists. There are well-documented changes which occur in the shape of a cell's nucleus in cancer. An *a priori* attempt to create a list of useful features for tumor/nontumor classification would thus most likely have included nuclear shape features. However, after our analysis, we find that the particular

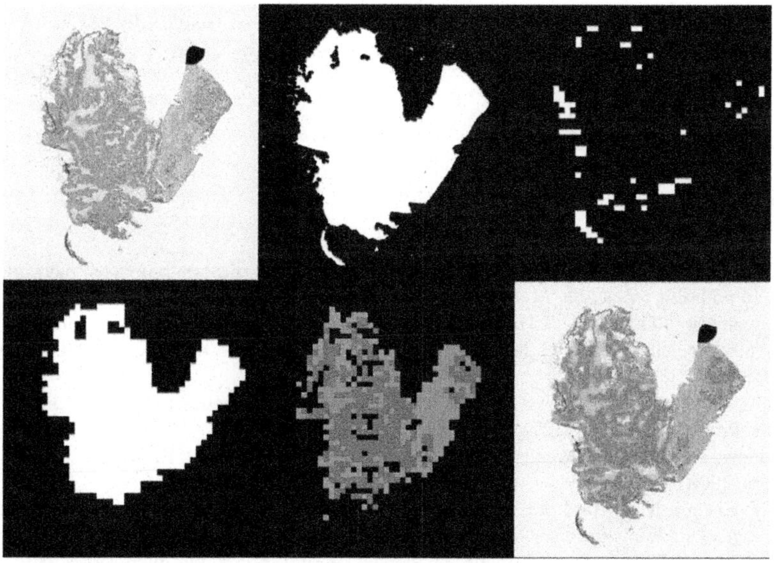

Fig. 4 (*Top-left*) A sample WSI. (*Top-center*) The region of interest, after removing background and pen marks. (*Top-right*) Tiles of the slide assessed as containing folded tissue, rendering them unsuitable for analysis. (*Bottom-left*) The final quality control results, indicating which image tiles are to be further processed. (*Bottom-center*) The predicted tissue type after applying the classifier. Tumor tissue is shown in green, nontumor tissue such as stroma and necrosis are shown as red. Black indicates a tile that was excluded by quality control. (*Bottom-right*) Classified tiles overlaid on the original WSI, showing the alignment of features between the two

metrics of nuclear shape examined were either not informative or redundant with other features, such as color, topology, and general basophilic object shape properties. This illustrates why it is valuable to start with a comprehensive feature set and narrow it empirically for each specific problem.

Figure 4 shows the results of the optimal tumor versus nontumor classification model (trained on all 512 × 512 segments) overlaid on the original WSI [68]. The tumor region (green) is cleanly separated from the stroma and necrotic tissue (red). Areas shown without an overlaid color are tiles that were excluded during quality control. All methods were implemented in MATLAB (Mathworks, Natick, MA).

Conclusion

This chapter highlights key challenges in the field of biomedical imaging informatics and suggests some techniques to overcome these challenges in histopathological images. Although these techniques perform decently well on smaller datasets, further research is needed to apply and validate these methods on larger datasets. With the development of The Cancer Genome Atlas, researchers now have free access to large public repositories. Therefore, the research community has to make an active effort to use this resource for thorough validation of their methodologies. In addition, an effort should be made to distribute the source code with publications, so that the community can easily use these methods to achieve their research goals.

References

1. M.Y. Gabril, G.M. Yousef, Informatics for practicing anatomical pathologists: marking a new era in pathology practice. Mod. Pathol. **23**, 349–358 (2010)
2. S. Kothari, J.H. Phan, T.H. Stokes, M.D. Wang, Pathology imaging informatics for quantitative analysis of whole-slide images. J. Am. Med. Inform. Assoc. **20**, 1099–1108 (2013)
3. J.R. Swedlow, I.G. Goldberg, K.W. Eliceiri, Bioimage informatics for experimental biology. Ann. Rev. Biophys. **38**, 327–346 (2009)
4. H. Peng, Bioimage informatics: a new area of engineering biology. Bioinformatics **24**, 1827–1836 (2008)
5. K.W. Eliceiri, M.R. Berthold, I.G. Goldberg, L. Ibáñez, B.S. Manjunath, M.E. Martone et al., Biological imaging software tools. Nat. Methods **9**, 697–710 (2012)
6. Z. Xiaobo, S.T.C. Wong, Informatics challenges of high-throughput microscopy. IEEE Signal Process. Mag. **23**, 63–72 (2006)
7. L.R. Long, S. Antani, T.M. Deserno, G.R. Thoma, Content-based image retrieval in medicine: retrospective assessment, state of the art, and future directions. Int. j healthc. inform. syst. inform. official publ. Inf. Res. Manag. Assoc. **4**, 1–16 (2009)
8. U. Sinha, A. Bui, R. Taira, J. Dionisio, C. Morioka, D. Johnson et al., A review of medical imaging informatics. Ann. N. Y. Acad. Sci. **980**, 168–197 (2002)

9. T. Liu, H. Peng, X. Zhou, Imaging informatics for personalised medicine: applications and challenges. Int. J. Funct. Inf. Personalised med. **2**, 125–135 (2009)
10. A. Wetzel, computational aspects of pathology image classification and retrieval. J. Supercomputing **11**, 279–293 (1997)
11. T.J. Fuchs, J.M. Buhmann, Computational pathology: challenges and promises for tissue analysis. Comput. Med. Imaging Graph. **35**, 515–530 (2011)
12. W. Amin, U. Chandran, V. Parwani Anil, J. Becich Michael, in *Essentials of Anatomic Pathology,* ed. by L. Cheng, D. G. Bostwick. Biomedical Informatics for Anatomic Pathology, (Springer, New York, 2011), pp. 469–480
13. M.N. Gurcan, L. Boucheron, A. Can, A. Madabhushi, N. Rajpoot, B. Yener, Histopathological image analysis: a review. IEEE Rev. Biomed. Eng. **2**, 147–171 (2009)
14. E.T. Sadimin, D.J. Foran, Pathology imaging informatics for clinical practice and investigative and translational research. North Am. J. Med. Sci. (Boston) **5**, 103–109 (2012)
15. L. Pantanowitz, P.N. Valenstein, A.J. Evans, K.J. Kaplan, J.D. Pfeifer, D.C. Wilbur et al., Review of the current state of whole slide imaging in pathology. J. Pathol. Inform. **2**, 36 (2011)
16. R. McLendon, A. Friedman, D. Bigner, E.G. Van Meir, D.J. Brat, G.M. Mastrogianakis et al., Comprehensive genomic characterization defines human glioblastoma genes and core pathways. Nature **455**, 1061–1068 (2008)
17. S. Kothari, J. Phan, T. Stokes, A. Osunkoya, A. Young, M. Wang, Removing batch effects from histopathological images for enhanced cancer diagnosis. IEEE J. Biomed. Health Inform. **18**, 765–772 (2014)
18. S. Kothari, J. Phan, M. Wang, Eliminating tissue-fold artifacts in histopathological whole-slide images for improved image-based prediction of cancer grade. J. Pathol. Inf. **4**, 22 (2013)
19. Y. Saeys, I. Inza, P. Larrañaga, A review of feature selection techniques in bioinformatics. Bioinformatics **23**, 2507–2517 (2007)
20. S. Palokangas, J. Selinummi, O. Yli-Harja, in *29th Annual International Conference of the IEEE Engineering in Medicine and Biology Society* Segmentation of folds in tissue section images, (2007), pp. 5642–5645
21. P. A. Bautista, Y. Yagi, in *Annual International Conference of the IEEE Engineering in Medicine and Biology Society*, Detection of tissue folds in whole slide images, (2009), pp. 3669–3672
22. F.W. Leong, M. Brady, J.O.D. McGee, Correction of uneven illumination (vignetting) in digital microscopy images. J. Clin. Pathol. **56**, 619–621 (2003)
23. S. Kothari, J. H. Phan, R. A. Moffitt, T. H. Stokes, S. E. Hassberger, Q. Chaudry, et al., in *IEEE International Symposium on Biomedical Imaging: From Nano to Macro*, Automatic batch-invariant color segmentation of histological cancer images, (2011), pp. 657–660
24. M. Macenko, M. Niethammer, J. S. Marron, D. Borland, J. T. Woosley, G. Xiaojun, et al., in *6th IEEE international conference on Symposium on Biomedical Imaging: From Nano to Macro,* A method for normalizing histology slides for quantitative analysis,(2009), pp. 1107–1110
25. D. Magee, D. Treanor, D. Crellin, M. Shires, K. Smith, K. Mohee, et al., in *Proc Optical Tissue Image analysis in Microscopy, Histopathology and Endoscopy (MICCAI Workshop),* Colour Normalisation in Digital Histopathology Images, (2009), pp. 100–111
26. K. Jun, H. Shimada, K. Boyer, J. Saltz, M. Gurcan, in *4th IEEE International Symposium on Biomedical Imaging: From Nano to Macro,* Image analysis for automated assessment of grade of neuroblastic differentiation, (2007), pp. 61–64
27. Q. Chaudry, S. Raza, A. Young, M. Wang, Automated renal cell carcinoma subtype classification using morphological, textural and wavelets based features. J. Sig Proc. Syst. **55**, 15–23 (2009)
28. C. Meurie, G. Lebrun, O. Lezoray, A. Elmoataz, A comparison of supervised pixels-based color image segmentation methods. application in cancerology. WSEAS Trans. Comput. **2**, 44–739 (2003)

29. K. Mao, P. Zhao, P. Tan, Supervised learning-based cell image segmentation for P53 immunohistochemistry. IEEE Trans. Biomed. Eng. **53**, 1153–1163 (2006)
30. P. Ranefalla, L. Egevadb, B. Nordina, E. Bengtssona, A new method for segmentation of colour images applied to immunohistochemically stained cell nuclei. Anal. Cell. Pathol. **15**, 145–156 (1997)
31. Y. Al-Kofahi, W. Lassoued, W. Lee, B. Roysam, Improved automatic detection and segmentation of cell nuclei in histopathology images. IEEE Trans. Biomed. Eng. **57**, 841–852 (2010)
32. O. Sertel, J. Kong, U. Catalyurek, G. Lozanski, J. Saltz, M. Gurcan, Histopathological image analysis using model-based intermediate representations and color texture: follicular lymphoma grading. J. Sig. Process. Syst. **55**, 169–183 (2009)
33. C. Gunduz-Demir, M. Kandemir, A. Tosun, C. Sokmensuer, Automatic segmentation of colon glands using object-graphs. Med. Image Anal. **14**, 1–12 (2010)
34. J.P. Monaco, J.E. Tomaszewski, M.D. Feldman, I. Hagemann, M. Moradi, P. Mousavi et al., High-throughput detection of prostate cancer in histological sections using probabilistic pairwise Markov models. Med. Image Anal. **14**, 617–629 (2010)
35. P. Thevenaz, M. Unser, Snakuscules. IEEE Trans. Image Process. **17**, 585–593 (2008)
36. H. Kong, M. Gurcan, K. Belkacem-Boussaid, Partitioning histopathological images: an integrated framework for supervised color-texture segmentation and cell splitting. IEEE Trans. Med. Imaging **30**, 1661–1677 (2011)
37. A. Tabesh, M. Teverovskiy, P. Ho-Yuen, V.P. Kumar, D. Verbel, A. Kotsianti et al., Multifeature prostate cancer diagnosis and gleason grading of histological images. IEEE Trans. Med. Imaging **26**, 1366–1378 (2007)
38. T. Fuchs, P. Wild, H. Moch, J. Buhmann, in *Medical Image Computing and Computer-Assisted Intervention*, Computational Pathology Analysis of Tissue Microarrays Predicts Survival of Renal Clear Cell Carcinoma Patients, (2008), pp. 1–8
39. M. Rahman, P. Bhattacharya, B.C. Desai, A framework for medical image retrieval using machine learning and statistical similarity matching techniques with relevance feedback. IEEE Trans. Inf Technol. Biomed. **11**, 58–69 (2007)
40. L. Yang, O. Tuzel, W. Chen, P. Meer, G. Salaru, L.A. Goodell et al., PathMiner: a web-based tool for computer-assisted diagnostics in pathology. IEEE Trans. Inf Technol. Biomed. **13**, 291–299 (2009)
41. V. Kovalev, A. Dmitruk, I. Safonau, M. Frydman, and S. Shelkovich, in *Computer Analysis of Images and Patterns,* A Method for Identification and Visualization of Histological Image Structures Relevant to the Cancer Patient Conditions, vol. 6854, ed. by P. Real, D. Diaz-Pernil, H. Molina-Abril, A. Berciano, W. Kropatsch (Springer Berlin/Heidelberg, 2011), pp. 460–468
42. J. Kong, O. Sertel, H. Shimada, K.L. Boyer, J.H. Saltz, M.N. Gurcan, Computer-aided evaluation of neuroblastoma on whole-slide histology images: classifying grade of neuroblastic differentiation. Pattern Recogn. **42**, 1080–1092 (2009)
43. M.E. Celebi, H.A. Kingravi, B. Uddin, H. Iyatomi, Y.A. Aslandogan, W.V. Stoecker et al., A methodological approach to the classification of dermoscopy images. Comput. Med. Imaging Graph. **31**, 362–373 (2007)
44. M. Muthu Rama Krishnan, M. Pal, R. R. Paul, C. Chakraborty, J. Chatterjee, and A. K. Ray, in *Journal of Medical Systems,* Computer Vision Approach to Morphometric Feature Analysis of Basal Cell Nuclei for Evaluating Malignant Potentiality of Oral Submucous Fibrosis, vol. 36 (2012), pp. 1746–1756
45. L.A.D. Cooper, K. Jun, D.A. Gutman, W. Fusheng, S.R. Cholleti, T.C. Pan et al., An integrative approach for in silico glioma research. IEEE Trans. Biomed. Eng. **57**, 2617–2621 (2010)
46. S. Doyle, M. Feldman, J. Tomaszewski, A. Madabhushi, A boosted bayesian multi-resolution classifier for prostate cancer detection from digitized needle biopsies. IEEE Trans. Biomed. Eng. **59**, 1205–1218 (2010)

47. P.W. Huang, C.H. Lee, Automatic classification for pathological prostate images based on fractal analysis. IEEE Trans. Med. Imaging **28**, 1037–1050 (2009)
48. K. Jafari-Khouzani, H. Soltanian-Zadeh, Multiwavelet grading of pathological images of prostate. IEEE Trans. Biomed. Eng. **50**, 697–704 (2003)
49. D. Zhang, G. Lu, Review of shape representation and description techniques. Pattern Recogn. **37**, 1 (2004)
50. L. Boucheron, Object-and spatial-level quantitative analysis of multispectral histopathology images for detection and characterization of cancer, Ph.D thesis, University of California, Santa Barbara, 2008
51. C. Gunduz, B. Yener, H.S. Gultekin, The cell graphs of cancer. Bioinformatics **20**, i145–i151 (2004)
52. C.C. Bilgin, P. Bullough, G.E. Plopper, B. Yener, ECM-aware cell-graph mining for bone tissue modeling and classification. Data Min. Knowl. Disc. **20**, 416–438 (2009)
53. A.N. Basavanhally, S. Ganesan, S. Agner, J.P. Monaco, M.D. Feldman, J.E. Tomaszewski et al., Computerized image-based detection and grading of lymphocytic infiltration in HER2 + breast cancer histopathology. IEEE Trans. Biomed. Eng. **57**, 642–653 (2010)
54. J. Sudbø, R. Marcelpoil, A. Reith, New algorithms based on the Voronoi Diagram applied in a pilot study on normal mucosa and carcinomas. Anal. Cell. Pathol. **21**, 71–86 (2000)
55. J. Sudbo, A. Bankfalvi, M. Bryne, R. Marcelpoil, M. Boysen, J. Piffko et al., Prognostic value of graph theory-based tissue architecture analysis in carcinomas of the tongue. Lab. Invest. **80**, 1881–1889 (2000)
56. I. Guyon, A. Elisseeff, An introduction to variable and feature selection. J. Mach. Learn. Res. **3**, 1157–1182 (2003)
57. M. A. Hall, "Correlation-based feature selection for machine learning," Department of Computer Science, Waikato University, New Zealand, 1999
58. C. Ding, H. Peng, Minimum redundancy feature selection from microarray gene expression data. J. Bioinf. Comput. Biol. **3**, 185–205 (2005)
59. I. Kononenko, in *Machine Learning: ECML-94,* ed. by F. Bergadano, L. De Raedt. Estimating attributes: Analysis and extensions of RELIEF, vol. 784 (Springer Berlin/Heidelberg, 1994), pp. 171–182
60. D. B. Skalak, in *Conference Processing on Machine Learning*, Prototype and Feature Selection by Sampling and Random Mutation Hill Climbing Algorithms, (1994), pp. 293–301
61. J.H. Holland, *Adaptation in natural and artificial systems: an introductory analysis with applications to biology, control, and artificial intelligence* (Michigan University, Ann Arbor, 1975)
62. S. Kirkpatrick, C.D. Gelatt, M.P. Vecchi, Optimization by simulated annealing. Science **220**, 671 (1983)
63. I. Guyon, J. Weston, S. Barnhill, V. Vapnik, Gene selection for cancer classification using support vector machines. Mach. Learn. **46**, 389–422 (2002)
64. S. B. Kotsiantis, in *Informatica (03505596),* Supervised machine learning: a review of classification techniques, vol. 31 (2007)
65. R. Bellazzi, B. Zupan, Predictive data mining in clinical medicine: current issues and guidelines. Int. J. Med. Inform. **77**, 81–97 (2008)
66. R. Hoffman, S. Kothari, J. Phan, M. D. Wang, in *The International Conference on Health Informatics*, ed. by Y.T. Zhang. A High-Resolution Tile-Based Approach for Classifying Biological Regions in Whole-Slide Histopathological Images, (Springer International Publishing, 2014), pp. 280–283
67. C.-C. Chang, C.-J. Lin, in *ACM Transactions on Intelligent Systems and Technology (TIST),* LIBSVM: a library for support vector machines, vol. 2 (2011), p. 27
68. S. Kothari, J. H. Phan, A. O. Osunkoya, M. D. Wang, in *Proceedings of the ACM Conference on Bioinformatics, Computational Biology and Biomedicine*, Biological interpretation of morphological patterns in histopathological whole-slide images, (2012), pp. 218–225

ECG Annotation and Diagnosis Classification Techniques

Yan Yan, Xingbin Qin and Lei Wang

Abstract ECG annotation has been studied for decades for the development of signal processing techniques and artificial intelligence methods. In this chapter, the general technique roadmaps of ECG beat annotation (classification) are reviewed. The deep neuro network methods are introduced after the mention of supervised and unsupervised learning methods as well as the deep belief networks. A preliminary study on deep learning application in ECG classification is proposed in this chapter, which leads to better results and has a high potential both for performance improvement and unsupervised learning applications.

Background

The heart is comprised of rhythmically contracting and thus drive the circulation of blood throughout the human body. A wave of electrical current passes through the entire heart, which triggers myocardial contraction [15]. Electrical propagation spreads over the whole heart in a coordinated pattern generate changes on the body surface potentials which can be measured and illustrated as an electrocardiogram (ECG, or sometimes EKG). Metabolic abnormalities (a lack of oxygen, or ischemia) and pathological changes of the heart engender the variety of ECG. Consequently, ECG analysis has been a routine of any complete medical evaluation or healthcare applications.

The automated ECG analysis provides primary assistance in clinical monitoring; a large number of approaches had been proposed for the task, basically the diag-

Y. Yan (✉) · X. Qin · L. Wang
Shenzhen Institutes of Advanced Technology, Chinese Academy of Sciences, Shenzhen, China
e-mail: yan.yan@siat.ac.cn

X. Qin
e-mail: xb.qin@siat.ac.cn

L. Wang
e-mail: wang.lei@siat.ac.cn

D. Xu et al. (eds.), *Health Informatics Data Analysis*, Health Information Science,
DOI 10.1007/978-3-319-44981-4_9

nosis of arrhythmia and further the inspection of heart rate variability or heart turbulence analysis [31]. Lots of ECG annotation and diagnosis classification techniques had been proposed in industrial circles and academic communities. The ECG classification includes data collection, preprocessing, feature extraction, and classification with a classifier.

Most of the literature described models combined with different classifier with features extracted by different feature extraction algorithms. The ECG classification methods develop at the same pace with the development of classification theories in machine learning and pattern recognition. In medical data collection and data annotation, ECG classification and detection forms similar research topics as speech recognition, natural language processing, and image processing.

In this chapter, we first introduce the basic elements and procedures in a typical ECG classification task, and then we would review shortly about the proposed literature of ECG classification, in the end, we would introduce new methods in unsupervised learning for ECG classification.

Technology Roadmap

ECG classification methods had been developed for decades. With the development of theories of machine learning and data mining, lots of algorithms had been adopted in this domain. Before the review of the methods, it is quite necessary to mention the typical experiment settings and data sets, as well as the framework of a classification problem which illustrated in Fig. 1.

ECG Acquisition

Acquiring and storing ECG data were the base for an analyzing task. Errors might creep into an analysis at any possible stage, not only the acquisition hardware system but also the transmission and storage should be carefully designed. The explanation for the acquisition field could be found in [14]. A raw data acquisition task related the digital signal processing and hardware design knowledge are out of the scope in this chapter; a typical ECG signal procurement process was illustrated in [45].

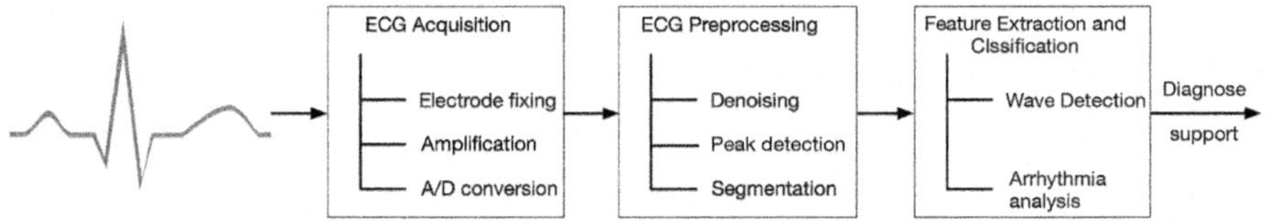

Fig. 1 The technology roadmap of an ECG classification task

As for the signal acquiring process, different kinds of sample rates might be involved, for conventional ECG acquisition device, the sample rate would be 128, 250, 340 or 500 Hz even higher. In murine studies, a sampling frequency of 2 kHz is considered sufficiently high [1]. Arbitrary resizing would be an ideal procedure to handle with the different sampling rate from the different data source to build the datasets for analysis.

ECG Signal Preprocessing

Before the segmentation and feature extraction process, the ECG signals were preprocessed. In ECG signals, the baseline wander (caused by Perspiration, respiration and body movements), power line interference and muscle noise were recorded as well, which had been described in lots of literature [9]. When the filtering methods were proposed and adopted in the preprocessing, the desired information should not be altered. The ECG typically exhibits persistent features like P-QRS-T morphology and average RR interval, and non-stationary features like individual RR and QT intervals, long-term heart rate trends [15]. Possible distortions caused by filtering should be quantified in these features.

The filtered ECG signals were segmented into individual heartbeat waveforms. ECG segmentation can be seen as the decoding procedure of an observation sequence regarding beat waveforms [2]. Dynamic time warping [51], time warping [50], Bayesian framework [41], hidden Markov models [2], weighted diagnostic distortion [54], morphology and heartbeat interval-based methods [17], and automatic and genetic methods [22] had been used in this sub-task. The state accuracies were close to 100%, which would be accurate enough in most online and offline applications.

ECG Feature Extraction and Classification

After segmentation for the ECG records, we got plenty of ECG waveform samples with variety categories. Since different physiological disorder might be reflected in different type of abnormal heartbeat rhythms. It is quite necessary to determine the classes. In the early literature, there were no unified class labels for an ECG classification problem. As in the MIT-BIH arrhythmia database annotations [32, 35], the class label system was build with five beat classes recommended by ANSI/AAMI EC57:1998 standard, i.e., normal beat, ventricular ectopic beat (VEB), supraventricular ectopic beat (SVEB), fusion of a normal and a VEB, or unknown beat type were used in most literature on the classification problems instead of early diversity subclass labels, which could beappropriate for the task since the widely acceptance.

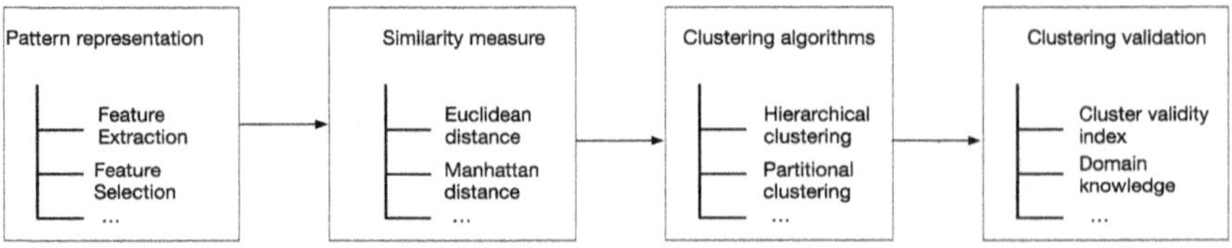

Fig. 2 An overview of the basic steps constituting a clustering process [15]

Supervised and Unsupervised Learning Methods in ECG Classification

The application of supervised learning methods had been widely used in ECG for the recognition and classification of different arrhythmia types. Lots of solutions have been proposed for the automated systems to annotate the ECG on real-time applications (e.g., [38, 41, 51]). Linear discriminate systems (e.g., [43], decision tree-based methods (e.g., [12, 28]), multilayer perceptron-based methods (e.g., [34]), fuzzy or neuro-fuzzy systems (e.g., [19, 48]), support vector machines classifiers (e.g., [47]), as well as the hybrid systems (e.g., [26, 49]) combined by those methods had been proposed. The details of these system are out of the scope of this chapter, later in the application sections, a comparison with some of the traditional methods had been reviewed in Fig. 2.

In addition to the supervised learning methods, unsupervised learning-based approaches became crucial in exploratory visualization-driven ECG analysis, which is useful for the detection of relevant trends, patterns, and outliers (e.g., [29, 44]). The most widely used methods for unsupervised learning in the recent research focused on the clustering-based techniques. Clustering-based methods learn the relevant similarity relationships of patterns which generate collections of clusters [7]. The clusters can be referred to a group of data vectors and then the similarities were calculated for the determination of class labels. Recently with the developing in neural networks techniques, deep learning-based methods would be introduced in this chapter, as the clustering methods had been described in the recent literature (e.g., [40, 53]).

Deep Learning in ECG Classification: A Preliminary Study-Based on Deep Sparse Autoencoder

Deep learning methods attempt to learn feature hierarchies as higher level features are formed by the composition of lower level features. The electrocardiography interpretation has been judged by the medical professionals, which was based on the abstractions of the perceptible features. In this model, we consider the higher level abstractions as the perceptible features, with whose composition the medical professionals can make arrhythmia judgement. The deep architecture automatic learning method is of particular importance for high-level abstractions, which human often do not know how to specify explicitly regarding raw sensory input [20]. As Collobert and Weston [16] discussed, deep learning methods are being used in learning internal representations of data. Another significant advantage they offer is the ability to naturally leverage: (a) unsupervised data and (b) data from similar tasks to boost performance on large and challenging problems that routinely suffer from a poverty of labeled data. In the electrocardiography classification problem, we got plenty of unsupervised data, and the labeled data was limited as well, so it is a free idea to adopt deep learning method to this problem.

Deep Neural Networks

The artificial neural network had been widely used in different applications, the basic 3-layer model (with only one hidden layer) is a relatively shallow network which means only shallow features can be learned via the structure. Deep neural networks were the structures in which we have multiple hidden layers, with which we can compute much more complex features from the input. Each hidden layer computes a nonlinear transformation of the previous layer, a deep network can have significantly greater representational power (i.e., can learn significantly more complex functions) than a shallow one. A typical deep neural network structure as in Fig. 3 makes no different from the normal multi-layer neural network.

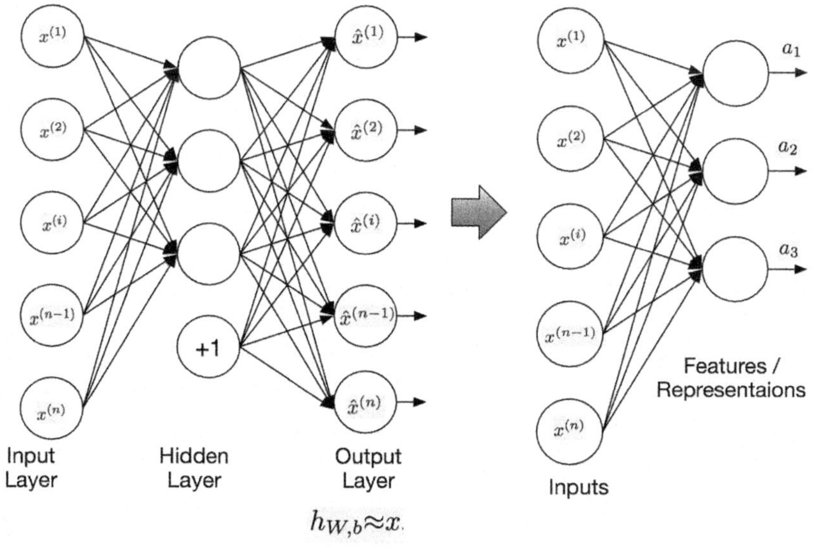

Fig. 3 A typical neural network structure

Autoencoders and Sparsity

An autoencoder is trained to encode the input x into some representation $c(x)$ inputs can be reconstructed from that representation. High-dimensional data can be converted to low-dimensional codes by training a multilayer neural network with a small central layer to reconstruct high-dimensional input vectors, and such "autoencoder" systems work better than principal components analysis as a tool to reduce the dimensionality of data [25]. Principal Component Analysis (PCA) is a linear reduction technique that seeks projection of the data into the directions of highest variability [18], while autoencoders do the same task in a different way with a wider scope (PCA is method that assumes linear systems whereas autoencoders do not). Since in the neural network the hidden layer is nonlinear, the autoencoder behaves differently from PCA, which can capture multi-modal aspects of the input distribution (the representation of the input). The related literature experiments reported in Bengio et al. [6] suggest that in practice when trained with stochastic gradient descent, nonlinear autoencoders with more hidden units than inputs (called overcomplete) yield useful representations (in the sense of classification error measured on a network taking this representation in input). A further defense of autoencoder can be accessed from Bengio [5]. As the theory illustrated, the electrocardiography signal representations can be learned via the autoencoder structure and learning algorithms.

The structures and learning algorithms used were represented in lots of literature [8, 18]. Here we impose a sparsity constraint on the hidden units to guarantee the representations expression ability. So for the neuron in the neuron network would be "active" if its output value is close to 1, or as being "inactive" if its output value is close to 0 due to the adopted sigmoid activation function. Here $a_j^{(2)}(x)$ denote the activation of hidden unit j in the autoencoder with the given input of x. Fatherly, let

$$\hat{\rho}_j = \frac{1}{m}\sum_{i=1}^{m}[a_j^{(2)}(x^{(i)})] \tag{1}$$

be the average activation of hidden unit j (averaged over the training set). Approximately enforce the constraint:

$$\hat{\rho}_j = \rho \tag{2}$$

where ρ is a sparsity parameter, typically a small value close to zero (such as $\rho = 0.05$), which means the average activation of each hidden neuron j to be close to zero (0.05 for instance).

The overall cost function of neural network is denoted by $J(W, b)$ which was defined by:

$$J(W,b) = \left[\frac{1}{m}\sum_{i=1}^{m}\left(\frac{1}{2}\|h_{W,b}(x^{(i)}) - y^{(i)}\|^2\right)\right] + \frac{\lambda}{2}\sum_{l=1}^{n_l-1}\sum_{i=1}^{s_l}\sum_{j=1}^{s_l+1} W_{ji}^{(l)2} \tag{3}$$

as the first term in the definition of $J(W,b)$ is an average sum-of-squares error term. The second term is a regularization term that tends to decrease the magnitude of the weights and helps prevent overfitting. The definition of λ, s, l, etc., would be explained in detail in the appendix part. To satisfy the constraint of sparsity, an additional penalty term to the optimisation objective that penalized $\hat{\rho}_j$ deviating significantly from ρ. The Kullback–Leibler (KL) divergence:

$$\sum_{j=1}^{s_2} KL(\rho||\hat{\rho}) = \sum_{j=1}^{s_2} \rho \log\frac{\rho}{\hat{\rho}_j} + (1-\rho)\log\frac{1-\rho}{1-\hat{\rho}_j} \tag{4}$$

is chosen as the penalty term. KL-divergence is a standard function for measuring how different two different distributions are. So in the autoencoder neural network training, the cost function of $J_{\text{sparse}}(W,b)$ was defined as:

$$J_{\text{sparse}}(W,b) = J(W,b) + \beta\sum_{j=1}^{s_2} KL(\rho||\hat{\rho}_j) \tag{5}$$

β denotes the weight of the sparsity penalty term. The above theories were cited from the recent research literature (e.g., [55]) and open source [36] on the topic of deep learning.

Representation Learning

The autoencoder base network had been used to learn representations (features) from unlabelled data. Autoencoders have been used as building blocks to build and initialize a deep multi-layer neural network. The training procedures are illustrated in Fig. 4 [5]:

1. Train the first layer as an autoencoder to minimize some form reconstruction error in the raw input. This is unsupervised.
2. The hidden units' outputs of the autoencoder are now used as input for another layer, also trained to be an autoencoder. Here unlabelled representations were used as well.
3. Iterates as in (2) to initialize the desired number of additional layers.
4. Take the last hidden-layer output as input to a supervised layer and initialize its parameters (either randomly or by supervised training, keeping the rest of the network fixed).

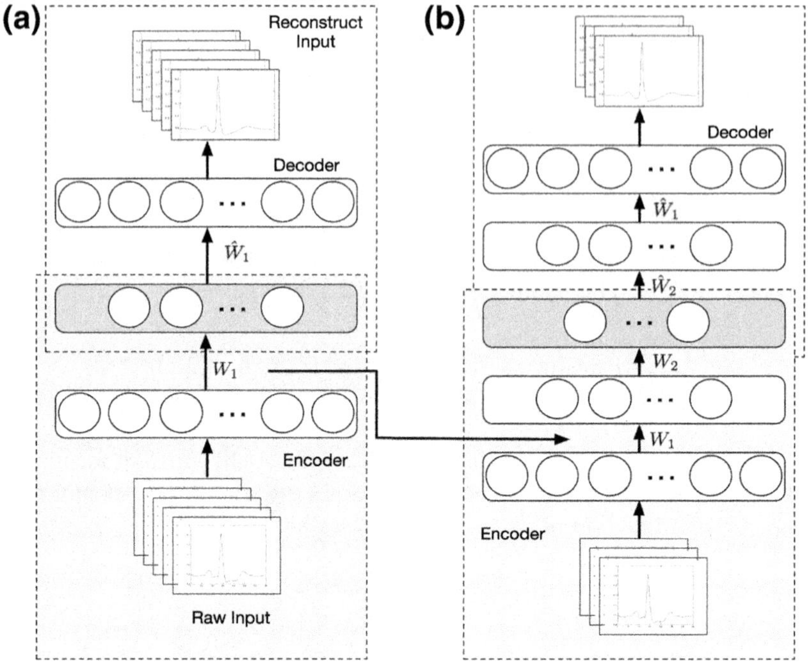

Fig. 4 A typical autoencoder neural network structure training process

5. Fine-tune all the parameters of this deep architecture on the supervised criterion. Alternately, unfold all the autoencoders into a very deep autoencoder and fine-tune the global reconstruction error.

The greedy layer-wise approach for pretraining a deep network works by training each layer in turn as explained in step (2). Assume $a^{(n)}$ as the deepest activation of the autoencoder network, then $a^{(n)}$ is a higher level representation than any lower layers, which contains what we interested. Then the higher level representations (the corresponding features in the traditional artificial selected features) can be used as the classifier input.

Fine-Tuning and Classifier

For training stacked autoencoders, when the parameters of one layer are being trained, parameters in other layer are kept fixed. Fine-tuning using backpropagation can be used to improve the model performance by tuning the parameters of all layers are changed at the same time after the layer-wise train phase. After the fine-tuning process, the optimized network structure would learn a good representation of the inputs, which can be used as the features similar to the traditional methods. The cardiac arrhythmia classification problem is a multi-class classification problem where the class label y may take more than two possible values. So the softmax regression is selected as the supervised learning algorithm which would be adopted as the classifier in conduction with the deep network.

Softmax regression model was generalized from the logistic regression. Similar to the logistic regression, the training set

$$\{(x^{(1)}, y^{(1)}), (x^{(2)}, y^{(2)}), \ldots, (x^{(m)}, y^{(m)})\} \tag{6}$$

of m labeled examples, the input features are $x^{(i)} \in \Re^{(n+1)}$ (with x_0 corresponding to there intercept term). The labels are denoted by

$$y^{(i)} \in \{1, 2, 3, \ldots, k\} \tag{7}$$

which means k classes. Given a test input x, the hypothesis to estimate the probability that $p(y = j|x)$ for each value of $j = 1, \ldots, k$. I.e., the probabilities of the class labels taking on the k different possible values are estimated.

$$h_\theta(x^{(i)}) = \begin{bmatrix} p(y^{(i)} = 1|x^{(i)}; \theta) \\ p(y^{(i)} = 2|x^{(i)}; \theta) \\ \vdots \\ p(y^{(i)} = k|x^{(i)}; \theta) \end{bmatrix} \tag{8}$$

$$= \frac{1}{\sum_{j=1}^{k} e^{\theta_j^T x^{(i)}}} \begin{bmatrix} e^{\theta_1^T x^{(i)}} \\ e^{\theta_2^T x^{(i)}} \\ \vdots \\ e^{\theta_k^T x^{(i)}} \end{bmatrix} \tag{9}$$

in which $\theta_1, \theta_2, \ldots, \theta_k \in \Re^{(n+1)}$ are parameters of the model. The term $\frac{1}{\sum_{j=1}^{k} e^{\theta_j^T x^{(i)}}}$ was normalizes the distribution, so that it sums to one.

The cost function adopted for softmax regression is:

$$J(\theta) = -\frac{1}{m}\left[\sum_{i=1}^{m}\sum_{j=1}^{k} 1\{y^{(i)} = j\} \log \frac{e^{\theta_j^T x^{(i)}}}{\sum_{l=1}^{k} e^{\theta_l^T x^{(i)}}}\right] \tag{10}$$

where $1\{\cdot\}$ is the indicator function. There is no known closed-form way to solve for the minimum of $J(\theta)$, and an iterative optimisation algorithm synch as gradient descent of L-BFGS could be used for the minimal value (some other iterative optimization algorithms were mentioned in Ngiam et al. [37]). So the cost function and iteration equations would be:

$$J(\theta) = -\frac{1}{m}\left[\sum_{i=1}^{m}\sum_{j=1}^{k} 1\{y^{(i)} = j\}\log\frac{e^{\theta_j^T x^{(i)}}}{\sum_{l=1}^{k} e^{\theta_l^T x^{(i)}}}\right] + \frac{\lambda}{2}\sum_{i=1}^{k}\sum_{j=0}^{n}\theta_{jk}^2 (\lambda > 0) \tag{11}$$

and

$$\nabla_{\theta_j} J(\theta) = -\frac{1}{m}\sum_{i=1}^{m}\left[x^{(i)}(1\{y^{(i)} = j\} - p(y^{(i)} = j|x^{(i)};\theta))\right] + \lambda\theta_j(\lambda > 0) \tag{12}$$

By minimizing $J(\theta)$ with respect to θ, the softmax regression classifier would work properly for the classification task.

Experiments and Results

Datasets Preparation

As illustrated in the above section, the preprocessing and segmentation had been described. In the preprocessing stage, filtering algorithms were adapted to remove the artifact signals from the ECG signal. The signals include baseline wander, power line interference, and high-frequency noise. The segmentation method based on the program of Laguna et al.[1] was adapted, which also had been validated by other related work [13] automatic. The experiment was based on three datasets:

1. Ambulatory electrocardiography database was used in this study, which includes recordings of 100 subjects with arrhythmia along with normal sinus rhythm. The database contains 100 records, each containing a $3 - lead$ $24 - hour$ long electrocardiography which was bandpass filtered at 0.1 – 100 Hz and sampled at 128 Hz. In this study, only the lead I data were adapted after preprocessing in the classification task. The average reference heart beats for each sample has 97,855 beats for the $24 - hour$ long recording, and the reference arrhythmia average is 1810 beats which were estimated by a commercial software (these statistics aim to indicate the existence of arrhythmia samples, which should not be considered as an experiment preset).
2. The MIT-BIH Arrhythmia Database [23] contains 48 half-hour recordings each containing two 30 min ECG lead signals (lead A and lead B), sampled at 360 Hz. As well only the lead I data were used in the proposed method. In agreement with the AAMI recommended practice, the four recordings with paced beats were removed from the analysis. Five records randomly selected were used to verify the real-time application. The remaining records were divided into two datasets, with the small part of which were used as the training set of the fine-tuning process (details would be described in the following part).

[1]"ecgpuwave", check the website of Physionet.

Table 1 Samples after Segementation

Ambulatory ECG Database (AECG)	MITBIH-AR	MITBIH-LT
9,785,500	100,687	667,343

Table 2 Samples dataset settings

Dataset	DS1	DS2	DS3
Useage	Pre-training	Fine-tuning	Test
Source (samples)	AECG (9,785,500)		
	AR (50,193)	AR (33,663)	AR (16,831)
	LT (587,347)	LT (50,000)	LT (30,000)
Total	10,423,040	83,633	46,831

3. The MIT-BIT Long-term Database is also used in this study for training and verification, which contains seven long-term ECG records (14–22 h each), with manually reviewed beat annotations and sampled at 128 Hz. Similarly, the seven recordings were divided into two datasets, with part used as the fine-tuning training set.

After the segmentation for the ambulatory ECG database, three batches of heartbeat samples listed in Table 1 were acquired for the classification task.

As for pretraining and fine-tuning stage for our proposed task and comparison, we divided all the samples into three groups: the pretraining group as DS1, the fine-tuning group as DS2 and test group as DS3 (illustrated in Table 2). Samples are chosen randomly from the original AR and LT database, the details of the sample class would be described in the experiment result analysis.

Classification Workflow

The stack autoencoder uses multilayer encoder network to transform high-dimensional data into low-dimensional code, similarly, a decoder network can be adopted to recover from the code, which we previously described. For the one-hidden-layer autoencoder input layer and hidden layer, the output was set equal to the input, starting with random weights in the one-hidden-layer neural networks, they can be trained together by minimizing the discrepancy between the original input data and the reconstruction. The gradients were obtained by using chain rule of backpropagation error derivatives; the decoder means the raw input data can be reconstructed by the learned feature with the trained weight. With large initial weights, autoencoders typically find poor local minima; with small initial weights, the gradients in the new layers are tiny, making it infeasible to train autoencoders with many hidden layers. After learning the feature and network weight in the first layer, we can add hidden layer one by one to get deeper representations, as well the learned weight can be used to reconstruct the input. When training the weight of layer 2, we take the fixed weights in layer one instead of random initialized weights because the learned weights are close to a good solution, which means training the parameters of each layer individually while freezing parameters for the remainder of the model. In the experiment, we adopted 2-hidden-layer, 3-hidden- layer, 4-hidden-layer stacked autoencoder for the test and verification.

Fine-tuning is a strategy that widely used in deep learning, which can be used to improve the performance of a stacked autoencoder. After pre-training multiple layers of feature detectors, the model is unfold to produce encoder and decoder networks that initially use the same weights. The weights learned can be used for classification implementation after adding one classifier after the feature layer. In this study, a softmax classifier was added. The parameters learned in the autoencoder pretraining were used in the fine-tuning initialization, and the weights W and biases b of softmax classifier (the last layer of the network) were initialized randomly. The training set of DS2 were used in the supervised learning pretraining while the backpropagation algorithm as usual of multi-layer perceptrons to minimize the output prediction error has been adopted.

The MIT-BIT Long-term Database is also used in this study for training and verification, which contains seven long-term ECG recordings (14–22 h each), with manually reviewed beat annotations and sampled at 128 Hz. Similarly, the seven records were divided into two datasets, with part used as the fine-tuning training set.

Classifier Performance Assessment

After the pretraining and fine-tuning process, the deep network parameters were acquired. The parameters and the test data set DS3 to predict the class of samples. It is necessary to mention that in DS2 and DS3, the labeled data used in pre-training and fine-tuning were divided randomly, which satisfy the requirement of Holdout cross-validation scheme so that the test results were meaningful for the classification task performance improvement.

The following statistical parameters of test performance were used in the study:

1. Specificity: number of correctly classified normal beats over total number of normal beats.
2. Sensitivity: number of correctly classified abnormal beats over total number of the given abnormal beats.
3. Overall classification accuracy: number of correctly classified beats over number of total beat.

Results

As previously mentioned, we adopted three different layer strategies for the classification task. In the 2-hidden-layer autoencoder network, we got a accuracy of 99.33%. For the N class the specificity is 99.76%, the sensitivity of S class is

80.08%, the sensitivity of V class is 98.13%, the sensitivity of F class is 85.48% as illustrated in Table 3.

In the 3-hidden-layer autoencoder network, we got a accuracy of 99.07%. For the N class the specificity is 99.64%, the sensitivity of S class is 75.14%, the sensitivity of V class is 97.58%, the sensitivity of F class is 80.33% as illustrated in Table 4.

In the 4-hidden-layer autoencoder network, we got a accuracy of 99.34%. For the N class the specificity is 99.74%, the sensitivity of S class is 82.29%, the sensitivity of V class is 98.31%, the sensitivity of F class is 87.71% as illustrated in Table 5.

Table 3 Test result for 2-hidden-layer autoencoder network

Algorithm classified label							
		N	S	V	F	Q	T
Reference label	N	41,965	39	45	13	6	42,068
	S	91	398	6	2	0	497
	V	63	3	3940	5	4	4015
	F	23	0	13	212	0	248
	Q	2	1	0	1	0	3

The test accuracy is about 99.33%

Table 4 Test results for 3-hidden-layer autoencoder network

Algorithm classified label							
		N	S	V	F	Q	T
Reference label	N	41,721	66	66	19	0	41,872
	S	120	405	13	1	0	539
	V	74	10	4073	17	0	4174
	F	27	2	19	196	0	244
	Q	2	0	0	1	0	2

The test accuracy is about 99.07%

Table 5 Test results for 4-hidden-layer autoencoder network

Algorithm classified label							
		N	S	V	F	Q	T
Reference label	N	41,778	38	48	17	5	41,886
	S	93	460	3	1	2	559
	V	52	1	4067	11	6	4137
	F	15	0	13	214	2	244
	Q	1	0	1	1	1	5

The test accuracy is about 99.34%

Table 6 Comparisons with other work using deep autoencoder

Approaches	Accuracy (%)	N-spe (%)	S-sen (%)	V-sen (%)	F-sen (%)
Proposed	**99.34**[b]	**99.76**	82.29	**98.31**	87.71
Mar et al. [31]	84.63	84.85	82.90	86.72	51.55
Chazal et al. [13]	86.19	86.86	**83.83**	77.74	**89.43**
Melgani and Bazi [33]	90.52	89.12	*[a]	89.97	*
Jiang and Kong [27]	94.51	98.73	50.59	86.61	35.78

[a]*Means the results were not available
[b]The listed percentages are based on the previous described rules Bold emphasized items are the highest score

Comparison with Other Work

Different kinds of performance assessment criteria had been adopted in the ECG arrhythmia classification problem. In the comparison part, we adopt several ordinary indicators for the performance assessment, which brought in the above sections. The accuracies, N-class specificities (N-spe), S-class sensitivities (S-sen), V-class sensitivities (V-sen), and the F-class sensitivities (F-sen) in Table 6 are presented for the comparison. The percentages are calculated from the literature' test results, in which some of the classes are ignored like melgan, we use a * symbol to represent the unavailable results. In Table 6, we use the highest value (2–4 hidden-layers structures, which are bold emphasized) for the verification which illustrated in "proposed" line.

Through the comparisons in Table 6, we can see that the proposed method offers better accuracy. Since accuracy in lots of the literature is good enough, the verification parameter depends on mainly on the normal class detection, but with these methods, this approach provided better performance in another kind of arrhythmia waveforms classes. Especially in the ventricular ectopic beat sensitivity, a quite considerable improvement had been made by the proposed method.

Deep Learning in ECG Classification: A Two-Lead ECG Classification Based on Deep Belief Network

A restricted Boltzmann machine learning algorithm was proposed in the two-lead heartbeat classification problem. A restricted Boltzmann machine (RBM) is a generative stochastic artificial neural network that can learn a probability distribution over its set of inputs [21]. In this part, a deep belief network was constructed, and the RBM-based algorithm was used in the classification problem.

The Deep Belief Network and Classifier

The Restricted Boltzmann Machine

The Restricted Boltzmann Machine is a stochastic neural network with substantial unsupervised learning ability. In the RBM network structure, each visible unit is connected to the hidden units without visible-visible or hidden-hidden connections. There are no links between the visible and hidden layers. The visible units are independent then the Gibbs sampling method could be used to approximate the probability distribution. It consists of one layer of visible units with input $X = (v_1, v_2, \ldots, v_n)$, one layer of hidden units with output $Y = (h_1, h_2, \ldots, h_m)$, and two bias units whose states are always on and a way to adjusting the value of each unit.

Boltzmann machine is based on statistical mechanics. The energy function $E(v, h)$ of an RBM is defined as:

$$E(v, h|\theta) = -\sum_{i=1}^{n} a_i v_i - \sum_{j=1}^{m} b_i h_i - \sum_{i=1}^{n}\sum_{j=1}^{m} v_i W_{ij} h_j \tag{13}$$

v and h present the state vectors of the visible and hidden layers, a_i. b_j and W_{ij} are parameters, define $\theta = \{W_{ij}, a_i, b_j\}$. So based on the energy function, the distribution of v and h is:

$$P(v, h|\theta) = \frac{e^{-E(v,h|\theta)}}{Z(\theta)}, Z(\theta) = \sum_{v,h} e^{-E(v,h|\theta)} \tag{14}$$

The purpose of RBM is to learn the optimal θ, according to the probability distribution, the maximum likelihood function is defined as:

$$\theta^* = arg \max_{\theta} L(\theta) = arg \max_{\theta} \sum_{t=1}^{T} \log P(v^{(t)}|\theta) \tag{15}$$

$$L(\theta) = \sum_{t=1}^{T} (\log \sum_{h} \exp[-E(v^{(t)}, h|\theta)] - \log \sum_{v}\sum_{h}[-E(v, h|\theta)]) \tag{16}$$

To get the optimal θ^*, stochastic gradient descent (e.g., [10]) method was used to maximum the likelihood function $L(\theta)$. The partial derivative of the parameters is shown below:

$$\frac{\partial \log P(v|\theta)}{\partial W_{ij}} = \langle v_i h_i \rangle_{\text{data}} - \langle v_i h_i \rangle_{\text{model}},$$
$$\frac{\partial \log P(v|\theta)}{\partial a_i} = \langle v_i \rangle_{\text{data}} - \langle h_i \rangle_{\text{model}}, \tag{17}$$
$$\frac{\partial \log P(v|\theta)}{\partial b_j} = \langle h_i \rangle_{\text{data}} - \langle h_i \rangle_{\text{model}}.$$

$\langle . \rangle_P$ denotes the distribution about P. $\langle . \rangle_{\text{data}}$ is easy to be calculated when the training samples were defined. $\langle . \rangle_{\text{model}}$ could not be resolved directly, but approximated by Gibbs sampling. Here we use the contrastive divergence algorithm (CD) [24] which would achieve better results by only one step of Gibbs sampling.

Classifier and the Training of Multi-layer RBM

This model generalized logistic regression [39] in classification missions which would be useful in heartbeats arrhythmia classification problems. The softmax model is a kind of supervised learning method in conjunction with the deep belief network.

Supposing m samples in the training set:

$$\{(x^{(1)}, y^{(1)}), (x^{(2)}, y^{(2)}), \ldots, (x^{(m)}, y^{(m)})\} \tag{18}$$

the inputs were vectors $x^{(i)}$ corresponding to the features space. The labels are denoted by $y^{(i)}$ corresponding to the arrhythmia classes of the inputs. The cost function of softmax regression with a weight decay term was defined as:

$$J(\theta) = -\frac{1}{m}\left[\sum_{i=1}^{m}\sum_{j=1}^{k} 1\{y^{(i)} = j\} \log \frac{e^{\theta_j^T x^{(i)}}}{\sum_{l=1}^{k} e^{\theta_l^T x^{(i)}}}\right] + \frac{\lambda}{2}\sum_{i=1}^{k}\sum_{j=0}^{n} \theta_{jk}^2 (\lambda > 0) \tag{19}$$

and the partial derivative of the parameters were:

$$\nabla_{\theta_j} J(\theta) = -\frac{1}{m}\sum_{i=1}^{m}[x^{(i)}(1\{y^{(i)} = j\} - p(y^{(i)} = j|x^{(i)}; \theta))] + \lambda\theta_j(\lambda > 0) \tag{20}$$

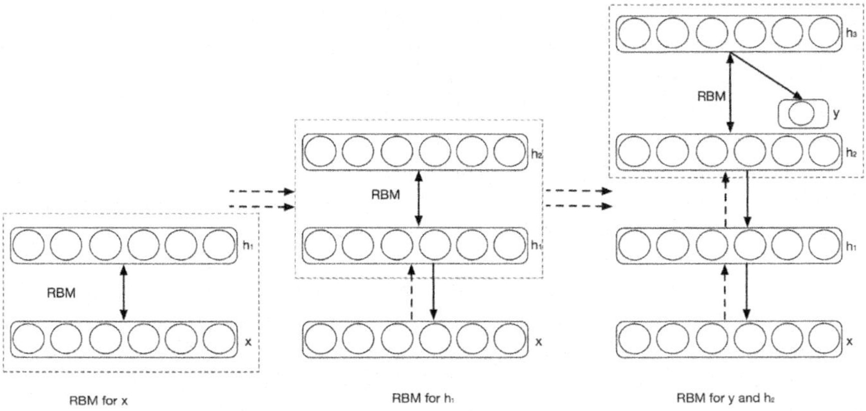

Fig. 5 The RBMs are stacked to from a deep belief network (DBN). The RBM can be trained layer by layer. It is easy to construct a DBN with the trained RBMs. Also, a softmax model to fine-tune all parameters behind the last layer

To train an optimum classifier, an optimization gradient descent algorithm called L-BFGS was used same as what had been done in the autoencoder classification.

Figure 5 shows that RBMs can be stacked and trained with a greedy manner to form a deep belief network(DBN) [5, 42]. In the last layer, a softmax classifier is connected with the DBN. DBN is the graphical model of a hierarchical architecture. The procedures were:

1. Train the first layer as an RBM which models the raw input X as a visible layer.
2. After training the RBM, representations of the input were obtained.
3. Train the next layer as an RBM which models the transformed data as a visible layer.
4. Iterate step 2 and 3 for the desired number of layers.

Finally, the RBMs are combined to a DBN with the softmax model. Fine-tuning then was used as the supervised method to minimize the likelihood function and improve the adaptability. Here the L-BFGS algorithm [3, 30] was used.

Combined Optimization Algorithm for Multi-lead Classifiers

The ECG wavelet transform performs differently in different channels by the waveforms. Each heartbeat channel shows diversely due to the P, QRS-complex and T-wave constituent. Then multi-lead ECG classification is significantly improved by the voting method. So a weight optimization method is proposed in two leads ECG signal classification. The method can be used in multi-lead ECG data classification.

For each classifier trained with distinct lead, a reliability value is denoted as the regular rate of the classifier. Let γ represent the reliability value. Then the classifier's reliability is defined as $\gamma_1, \gamma_2, \cdots$

In the matrix:

$$ClassSTST = \begin{pmatrix} C_{11} & C_{12} & \cdots & C_{1n} \\ C_{21} & C_{12} & \cdots & C_{2n} \\ \cdots & \cdots & \ddots & \cdots \\ C_{n1} & C_{n2} & \cdots & C_{nn} \end{pmatrix} \tag{21}$$

there are n classes and C_{11} represents the class 1 which is classified as class 1, C_{12} represents class 1 is classified as 2, etc. The diagonal values are the correct classification. The purpose is to increase the diagonal values, so we adopt weights of the outputs for each classifier. The weights of the first classifier is $W_1 = (w_{11}, w_{12}, \cdots, w_{1n})$, the second were $W_2 = (w_{21}, w_{22}, \cdots, w_{2n})$. The constraint condition is $\sum_{i=1}^{2} w_{ik} = 1$.

First initial the weight to a mean value. Each sample has an output vector $O = o_1, o_2, \ldots, o_n$, the class is decided by the maximum value. Adding the weight of the value, the output is $(o_1 w_1, o_2 w_2, \cdots, o_n w_n)$. If the label of the sample is l, we would like to maximize $o_1 w_{11} + o_1 w_{21}$ while correspondingly minimize others. Through this, the l class accurate is promoted while the false negative rate is also increased. The optimization algorithm would find a balance between the two weights of all testing samples. Accordingly, the optimal function is defined as:

$$F(x) = \frac{1}{\sqrt{2\pi\sigma}} x \exp^{-\frac{(x-\mu)^2}{2\sigma^2}} \tag{22}$$

To find the maximum value of x, the derivative of the equation is:

$$F'(x) = 0, x = \frac{\mu}{2} + \sqrt{\frac{\mu^2}{4} + \sigma^2} \tag{23}$$

Here, μ is denoted as initial weights. On the basis of the statistical matrix, counting the difference of the correct (diffcort) and false negative (diffflneg) quantities of the two classifiers. If the difference of the difference of the correct and false negative greater than zero, $\sigma^2 = \sqrt{\frac{\text{diffcort}-\text{diffflsneg}}{\text{total number}}}$, so updating the corresponding class weight of the first classifier as $w_1 = \frac{\mu}{2} + \sqrt{\frac{\mu^2}{4} + \sigma^2}$. Else, $\sigma^2 = \sqrt{\frac{\text{diffflsneg}-\text{diffcort}}{\text{total number}}}$, updating the weight of the second classifier as $w_2 = \frac{\mu}{2} + \sqrt{\frac{\mu^2}{4} + \sigma^2}$. Finally normalize the weights, the optimal combined value is $\gamma_1 o_1 w_1 + \gamma_2 o_2 w_2$. Generally, every two classifiers can be used to optimize the multi-lead ECG classification.

Experiment and Results

Preprocessing and Segmentation

Here we do the same process as before. The preprocessing include two main parts: the ECG data filtering and heartbeat segmentation. The filtering task is removing the artifact signal from the ECG signal, which includes the baseline wander, power line interference, and high-frequency noise. The massive unlabelled data we collected is extracted and resampled from 128 to 360 Hz.

In heartbeat segmentation process, average samples of each beat are 277 samples. To get more information, we allow partially overlap, and a window with a length of 340 data points in one beat was defined, the R peak of the wave is located at 141st point. Most annotations of the MIT-BIH arrhythmia database lied under the R-wave. For the dataset we collect, an high-accuracy algorithm has been explored to determine the R pick and then divide into the heartbeat segments according to the R pick.

Training and Fine-Tuning

The goal of RBM learning is to maximize the product of the probabilities. The parameters of the network can be initialized by the constructor. This option is useful when an RBM is used as the building block of the deep belief network, in which case the weight matrix and the hidden layer bias is shared with the corresponding layer of the network. The active function of the nodes is the sigmoid function. The data is batched to train the RBM layer by layer. A single-step contrastive divergence(CD-1) [11] is used in the gradient descent procedure. After calculating the partial derivative, the weights and bias are updated.

After the learning process, the RMBs can be used to initialize a deep belief network. Standard backpropagation algorithm can be applied to fine-tune the model. That can significantly improve the performance of the DBN. The fine-turning process is a supervised learning procedure, so at the last layer, a multi-class model called softmax is connected to classify the ECG data. Then using the fine-tuning method to minimize the cost function.

Experiment Results

In the experiments, we adopted multi-lead ECG signal based on the restricted Boltzmann machine for the classification task. The MIT-BIH arrhythmia database is divided into three parts. Half of the beat is added to training the RBMs; one-third is applied to fine-tune the network, and the left is used to test the model. In the three hidden layers deep belief network, the classifier outperforms regarding sensitivity (SPR) 99.35%, specificity (SPC) 95.18%, and accuracy rate (ACC) 98.25% using

Table 7 Test result of three hidden Layers deep belief network using the first lead

Algorithm classified label													
		N*[a]	L*	R*	AB*	P*	FU*	NPC	APC	FL*	V*	NE*	AE*
Original label	NORMAL	12,289	3	2	3	15	9	0	37	2	0	0	0
	LBBB	7	1369	0	0	6	0	0	1	0	0	0	0
	RBBB	4	0	1179	0	3	0	0	4	0	0	0	0
	ABERR	7	1	0	12	5	0	0	0	1	0	0	0
	PVC	18	2	0	2	1104	8	0	2	2	0	0	0
	FUSION	19	0	0	1	6	97	0	0	0	0	2	0
	NPC	5	0	1	0	0	0	4	1	0	0	1	0
	APC	58	2	9	0	2	0	0	382	0	0	2	0
	FLWAV	10	0	0	1	8	0	0	0	58	0	0	0
	VESC	2	0	0	0	1	0	0	0	0	24	0	0
	NESC	6	0	1	0	0	0	0	1	0	0	15	0
	AESC	3	0	0	0	0	0	0	0	0	0	0	0

The test accuracy of the first lead is 98.247%

[a]Use N* as NORMAL for abbreviation and the same for the rest symbols

Table 8 Test result of three hidden Layers deep belief network using the first lead

Algorithm classified label		N*	L*	R*	AB*	P*	FU*	NPC	APC	FL*	V*	NE*	AE*
Original label	NORMAL	12,233	7	3	3	57	13	1	34	12	0	6	0
	LBBB	12	1366	0	0	5	0	0	0	0	0	0	0
	RBBB	5	0	1169	0	5	0	0	9	2	0	0	0
	ABERR	7	0	1	10	6	0	1	0	1	0	0	0
	PVC	56	3	1	0	1048	15	0	4	11	0	0	0
	FUSION	22	0	0	0	5	97	0	0	0	0	1	0
	NPC	1	0	1	0	0	0	10	0	0	0	0	0
	APC	60	4	8	0	11	2	0	368	1	0	0	1
	FLWAV	7	0	0	0	8	0	0	1	61	0	0	0
	VESC	2	0	1	0	2	0	0	0	2	20	0	0
	NESC	6	0	1	0	1	0	0	0	0	0	15	0
	AESC	2	0	0	0	0	0	0	0	0	0	0	1

The test accuracy of the second is 97.433%

Table 9 Test result of three hidden Layers deep belief network using the first lead

Algorithm classified label													
		N*	L*	R*	AB*	P*	FU*	NPC	APC	FL*	V*	NE*	AE*
Original label	NORMAL	12,348	0	0	0	11	2	0	7	1	0	0	0
	LBBB	3	1377	0	0	3	0	0	0	0	0	0	0
	RBBB	1	0	1182	0	2	0	0	5	0	0	0	0
	ABERR	8	0	0	16	2	0	0	0	0	0	0	0
	PVC	16	0	0	0	1108	10	0	1	3	0	0	0
	FUSION	22	0	0	0	4	99	0	0	0	0	0	0
	NPC	3	0	0	0	0	0	9	0	0	0	0	0
	APC	58	2	5	0	1	0	0	389	0	0	2	0
	FLWAV	2	0	0	0	6	0	0	0	69	0	0	0
	VESC	3	0	0	0	1	0	0	0	0	23	0	0
	NESC	11	0	1	0	0	0	0	1	0	0	11	0
	AESC	3	0	0	0	0	0	0	0	0	0	0	0

The test accuracy of the second is 98.829%

Table 10 Comparisons with others' works

Approaches	Accuracy (ACC) (%)	Sensitivity (TPR) (%)	Specificity (SPC) (%)
Proposed	98.83	99.83	96.05
Tadejko and Rakowki [46]	97.82	99.70	93.10
Banerjee and Mitra [4]	97.60	97.30	98.80
Ye et al. [52]	99.71	*NR[a]	*NR
Osowski and Linh [38]	96.06	98.10	95.53

[a]*NR means the results were not reported
The listed percentages are based on the assessment rules

the first channels. Using the second channel, we get the accuracy (ACC) of 97.43%, sensitivity (SPR) 98.90% and specificity (SPC) 93.36%. At the convergence of the optimization process, the combining method achieves the accuracy (ACC) of 98.83%, sensitivity (SPR) 99.83%, and specificity (SPC) 96.05% (Table 7).

For we collect a large amount of ECG data from the hospital which contains lots of regular beats and abnormal beats, the learning method of restricted Boltzmann machine is used to learning the features from the massive data in an unsupervised way. Then the RBMs is adapted to build a deep belief network. The optimization algorithm we propose improves the accuracy of multi-lead ECG signal. In this comparison, we evaluate the performance on the indicators which put forward in the above sections: Sensitivity(TPR), specificity(SPC), and overall Accuracy(ACC) with the SVM, ICA, ANN methods. The $*\ NR$ symbol represents the result which were not reported as illustrated in Table 8 and Table 9.

Table 10 shows the performance of different models that used for the heartbeat classification, the proposed method offers a high accuracy of classification. All the annotations in the MIT-BIH arrhythmia database are used in our study and Osowski and Linh [38] only selects seven types. Ye et al. [52] get the highest accuracy but with a price of rejecting 2054 heartbeats (2.4% rejections). By comparing with others' experience, our approach provided higher performance in heartbeat classification without complex wavelet transform algorithms.

Conclusion

ECG annotation research had been developed for decades, the signal processing methods, feature extraction, and classifier had been studied diffusely. In this paper, we first reviewed the technique roadmap for an ECG classification task, which composed most of the ECG classification research literature. Then we make a summary on the classification methodology including supervised learning and unsupervised learning, which included most ECG classification methods. Then two kinds of new methods in unsupervised learning had been proposed for ECG annotations, which improve the state-of-art in accuracy and special arrhythmia beat detection rate.

Since the deep network autoencoder structure and deep belief network with the training algorithms had been widely used in modern computing science, an intuitionistic idea is an application in great ECG records. This study proposed one possibility to adopt this method in the health informatics Big Data applications. In the both structures we get higher performance than the recent research in this domain. As the evolvability of the system, the performance could be improved as the dataset grows, from which a new possibility to make use of the considerable amount of unlabelled ECG data from the long-term clinical monitoring and healthcare monitoring had been proposed. Since pre-training and fine-tuning bring the systems with the ability of self-learning, the structures could be better optimized as the training samples become larger, especially for the rare abnormalities (the S, V, F class, check Table 6). In the traditional literature, the MIT or AHA datasets were used which limited the samples of the abnormalities, so there were no good sensitivities, to the contrary using the unlabelled data and the self-learning ability the system could be improved in these outlier detections.

References

1. H. Ai, X. Cui, L. Tang, W. Zhu, X. Ning, X. Yang, Studies on the time domain and power spectrum of high frequency ecg in normal mice. Sheng li xue bao: [Acta Physiol. Sin.] **48**(5), 512–516 (1996)
2. R.V. Andreao, B. Dorizzi, J. Boudy, ECG signal analysis through hidden markov models. IEEE Trans. Biomed. Eng. **53**(8), 1541–1549 (2006)
3. G. Andrew, J. Gao, Scalable training of l 1-regularized log-linear models, in *Proceedings of the 24th International Conference on Machine learning* (ACM, 2007), pp. 33–40
4. S. Banerjee, M. Mitra, Application of cross wavelet transform for ECG pattern analysis and classification. IEEE Trans. Instrum. Meas. **63**(2), 326–333 (2014)
5. Y. Bengio, Learning deep architectures for AI. Foundations and Trends®. Mach. Learn. **2**(1), 1–127 (2009)
6. Y. Bengio, P. Lamblin, D. Popovici, H. Larochelle et al., Greedy layer-wise training of deep networks. Adv. Neural. Inf. Process. Syst. **19**, 153 (2007)
7. J.C. Bezdek, N.R. Pal, Some new indexes of cluster validity. Syst. Man Cybern. Part B IEEE Trans. Cybern. **28**(3), 301–315 (1998)
8. C.M. Bishop et al., *Pattern recognition and machine learning*, vol. 4 (Springer, New York, 2006)
9. M. Blanco-Velasco, B. Weng, K.E. Barner, Ecg signal denoising and baseline wander correction based on the empirical mode decomposition. Comput. Biol. Med. **38**(1), 1–13 (2008)
10. L. Bottou, Large-scale machine learning with stochastic gradient descent, in Proceedings of COMPSTAT'2010 (Springer, 2010), pp. 177–186
11. M.A. Carreira-Perpinan, G.E. Hinton, On contrastive divergence learning, in *Proceedings of the Tenth International Workshop on Artificial Intelligence and Statistics* (Citeseer, 2005), pp. 33–40
12. F. Charfi, A. Kraiem, Comparative study of ecg classification performance using decision tree algorithms. Int. J. E-Health Med. Commun. (IJEHMC) **3**(4), 102–120 (2012)

13. P. Chazal, M. O'Dwyer, R. Reilly, Automatic classification of heartbeats using ecg morphology and heartbeat interval features. IEEE Trans. Biomed. Eng. **51**(7), 1196–1206 (2004)
14. G.D. Clifford, F. Azuaje, P.E. McScharry, Advanced tools for ECG analysis (2006a). http://www.ecgtools.org/
15. G.D. Clifford, F. Azuaje, P. McSharry et al., *Advanced methods and tools for ECG data analysis* (Artech House, Boston, 2006)
16. R. Collobert, J. Weston, A unified architecture for natural language processing: deep neural networks with multitask learning, in *Proceedings of the 25th International Conference on Machine Learning* (ACM, 2008), pp. 160–167
17. P. De Chazal, M. O'Dwyer, R.B. Reilly, Automatic classification of heartbeats using ecg morphology and heartbeat interval features. IEEE Trans. Biomed. Eng. **51**(7), 1196–1206 (2004)
18. R.O. Duda, P.E. Hart, D.G. Stork, Pattern classification. Wiley (2012)
19. M. Engin, ECG beat classification using neuro-fuzzy network. Pattern Recogn. Lett. **25**(15), 1715–1722 (2004)
20. D. Erhan, P.A. Manzagol, Y. Bengio, S. Bengio, P. Vincent, The difficulty of training deep architectures and the effect of unsupervised pre-training, in *International Conference on Artificial Intelligence and Statistics*, pp. 153–160 (2009)
21. A. Fischer, C. Igel, An introduction to restricted boltzmann machines, in *Progress in Pattern Recognition, Image Analysis, Computer Vision, and Applications* (Springer, 2012), pp. 14–36
22. A. Gacek, W. Pedrycz, A genetic segmentation of ECG signals. IEEE Trans. Biomed. Eng. **50** (10), 1203–1208 (2003)
23. A.L. Goldberger, L.A.N. Amaral, L. Glass, J.M. Hausdorff, P.C. Ivanov, R.G. Mark, J.E. Mietus, G.B. Moody, C.K. Peng, H.E. Stanley, PhysioBank, PhysioToolkit, and PhysioNet: Components of a new research resource for complex physiologic signals. Circulation **101**(23), e215–e220 (2000)
24. G. Hinton, Training products of experts by minimizing contrastive divergence. Neural Comput. **14**(8), 1771–1800 (2002)
25. G.E. Hinton, R.R. Salakhutdinov, Reducing the dimensionality of data with neural networks. Science **313**(5786), 504–507 (2006)
26. M.R. Homaeinezhad, S. Atyabi, E. Tavakkoli, H.N. Toosi, A. Ghaffari, R. Ebrahimpour, ECG arrhythmia recognition via a neuro-svm–knn hybrid classifier with virtual QRS image-based geometrical features. Expert Syst. Appl. **39**(2), 2047–2058 (2012)
27. W. Jiang, S. Kong, Block-based neural networks for personalized ecg signal classification. IEEE Trans. Neural Networks **18**(6), 1750–1761 (2007)
28. V. Krasteva, R. Leber, I. Jekova, R. Schmid, R. Abacherli, Classification of supraventricular and ventricular beats by QRS template matching and decision tree, in *Computing in Cardiology Conference (CinC)* (IEEE, 2014), pp. 349–352
29. M. Lagerholm, C. Peterson, G. Braccini, L. Edenbrandt, L. S¨ornmo, Clustering ECG complexes using hermite functions and self-organizing maps. IEEE Trans. Biomed. Eng. **47** (7), 838–848 (2000)
30. D.C. Liu, J. Nocedal, On the limited memory BFGS method for large scale optimization. Math. Program. **45**(1–3), 503–528 (1989)
31. T. Mar, S. Zaunseder, J.P. Martinez, M. Llamedo, R. Poll, Optimization of ECG classification by means of feature selection. IEEE Trans. Biomed. Eng. **58**(8), 2168–2177 (2011)
32. R. Mark, P. Schluter, G. Moody, P. Devlin, D. Chernoff, An annotated ecg database for evaluating arrhythmia detectors. IEEE Trans. Biomed. Eng. **29**, 600–600 (1982)
33. F. Melgani, Y. Bazi, Classification of electrocardiogram signals with support vector machines and particle swarm optimization. IEEE Trans. Inf. Technol. Biomed. **12**(5), 667–677 (2008). doi:10.1109/TITB.2008.923147
34. K. Minami, H. Nakajima, T. Toyoshima, Real-time discrimination of ventricular tachyarrhythmia with fourier-transform neural network. IEEE Trans. Biomed. Eng. **46**(2), 179–185 (1999)

35. G.B. Moody, R.G. Mark, The mit-bih arrhythmia database on CD-ROM and software for use with it, in *Proceedings Computers in Cardiology*, pp. 185–188 (1990)
36. A. Ng, J. Ngiam, C.Y. Foo, Y. Mai, C. Suen, UFLDL Tutorial (2010). http://ufldl.stanford.edu/wiki/index.php/UFLDL Tutorial
37. J. Ngiam, A. Coates, A. Lahiri, B. Prochnow, Q.V. Le, A.Y. Ng, On optimization methods for deep learning, in *Proceedings of the 28th International Conference on Machine Learning (ICML-11)*, pp 265–272 (2011)
38. S. Osowski, T.H. Linh, ECG beat recognition using fuzzy hybrid neural network. IEEE Trans. Biomed. Eng. **48**(11), 1265–1271 (2001)
39. C.Y.J. Peng, K.L. Lee, G.M. Ingersoll, An introduction to logistic regression analysis and reporting. J. Educ. Res. **96**(1), 3–14 (2002)
40. J.L. Rodriguez-Sotelo, D. Peluffo-Ordonez, D. Cuesta-Frau, G. Castellanos-Dominguez, Unsupervised feature relevance analysis applied to improve ECG heartbeat clustering. Comput. Methods Programs Biomed. **108**(1), 250–261 (2012)
41. O. Sayadi, M. Shamsollahi, A model-based bayesian framework for ECG beat segmentation. Physiol. Meas. **30**(3), 335 (2009)
42. J. Schmidhuber, Deep learning in neural networks: an overview. Neural Networks **61**, 85–117 (2015)
43. M.F. Shinwari, N. Ahmed, H. Humayun, I. ul Haq, S. Haider, A. ul Anam, Classification algorithm for feature extraction using linear discriminant analysis and cross-correlation on ECG signals. Int. J. Adv. Sci. Technol. **48**, 149–161 (2012)
44. R. Silipo, G. Bortolan, C. Marchesi, Supervised and unsupervised learning for diagnostic ECG classification, in: Engineering in Medicine and Biology Society, 1996. Bridging Disciplines for Biomedicine. Proceedings of the 18th Annual International Conference of the IEEE, vol. 3 (IEEE, 1996), pp. 931–932
45. C.V. Silva, A. Philominraj, C. del Rio, A DSP Practical Application: Working on ECG Signal. INTECH Open Access Publisher (2011)
46. P. Tadejko, W. Rakowski, Hybrid wavelet-mathematical morphology feature extraction for heartbeat classification, in *EUROCON, 2007. The International Conference on "Computer as a Tool"* (IEEE, 2007), pp. 127–132
47. E.D. Übeyli, ECG beats classification using multiclass support vector machines with error correcting output codes. Digit. Signal Process. **17**(3), 675–684 (2007)
48. M. Vafaie, M. Ataei, H. Koofigar, Heart diseases prediction based on ECG signals classification using a genetic-fuzzy system and dynamical model of ECG signals. Biomed. Signal Process. Control **14**, 291–296 (2014)
49. L. Vanitha, G. Suresh, Hybrid svm classification technique to detect mental stress in human beings using ECG signals. In: *2013 International Conference on Advanced Computing and Communication Systems (ICACCS)* (IEEE, 2013), pp. 1–6
50. H. Vullings, M. Verhaegen, H.B. Verbruggen, ECG segmentation using timewarping, in *Advances in Intelligent Data Analysis Reasoning about Data* (Springer, 1997), pp. 275–285
51. H. Vullings, M. Verhaegen, H. Verbruggen, Automated ECG segmentation with dynamic time warping. In: *Engineering in Medicine and Biology Society, 1998. Proceedings of the 20th Annual International Conference of the IEEE* (IEEE, 1998), pp. 163–166
52. C. Ye, B.V. Kumar, M.T. Coimbra, Heartbeat classification using morphological and dynamic features of ECG signals. IEEE Trans. Biomed. Eng. **59**(10), 2930–2941 (2012)
53. Y.C. Yeh, C.W. Chiou, H.J. Lin, Analyzing ECG for cardiac arrhythmia using cluster analysis. Expert Syst. Appl. **39**(1), 1000–1010 (2012)
54. Y. Zigel, A. Cohen, A. Katz, The weighted diagnostic distortion (WDD) measure for ECG signal compression. IEEE Trans. Biomed. Eng. **47**(11), 1422–1430 (2000)
55. W. Zou, S. Zhu, K. Yu, A.Y. Ng, Deep learning of invariant features via simulated fixations in video, in *Advances in Neural Information Processing Systems*, pp. 3212–3220 (2012)

EEG Visualization and Analysis Techniques

Gregor Schreiber, Hong Lin, Jonathan Garza, Yuntian Zhang and Minghao Yang

Abstract We present some information depicting the current status of EEG research with projected applications in the areas of health care. We describe a method of quick prototyping an EEG headset, in a cost-effective way and with state-of-the-art technologies. We use meditation research to reach out to the high-end applications of EEG data analysis in understanding human brain states and assisting in promoting human health care. Some devotees to the practices of transcendental meditation have shown the ability to control these brain states. We want to numerically prove or disprove this assumption; the analysis of these states could be the initial step in a process to first predict and later allow individuals to control these states. To this end, we begin to build a system for dynamic and onsite brain state analysis using EEG data. The system will allow users to transit EEG data to an online database through mobile devices, interact with the web server through web interface, and get feedback from EEG data analysis programs on real-time bases.

G. Schreiber
Chevron-Phillips Chemical Company, Houston, TX, USA
e-mail: schreg@cpchem.com

H. Lin (✉) · J. Garza · Y. Zhang · M. Yang
Department of Computer Science and Engineering Technology,
University of Houston-Downtown, Houston, TX, USA
e-mail: linh@uhd.edu

J. Garza
e-mail: garzaj79@gator.uhd.edu

Y. Zhang
e-mail: yuntian.zh@gmail.com

M. Yang
e-mail: yangminghao@gmail.com

D. Xu et al. (eds.), *Health Informatics Data Analysis*, Health Information Science,
DOI 10.1007/978-3-319-44981-4_10

Background

Using electroencephalographic (EEG) data, cognitive psychologists can visualize and observe correlations between different active brain states. It is desirable to create an application that takes EEG data and exposes it to various analytical techniques so the resultant brain states can be studied and predicted. We present explanations of the design and implementation offered herein. The presentation will consist of an extrication of the design of an EEG headset, which can collect EEG, pulse, and temperature data, and a case study in which EEG signals demonstrate differences between different brain states.

An EEG device can record the electric signals from a human scalp. EEG devices used to be only available in professional healthcare institutions for clinic use. Last decade witnessed the development of cheap EEG devices, for example, EPOC from Emotiv (http://www.emotiv.com) and NeuroSky (http://www.neurosky.com/), and increasing interest in EEG-based brain–computer interfaces (BCI). EEG signals characterize the result of the neuron activities inside a human brain. Naturally, they are used to study and understand human brain activities. In particular, EEG signals indicate that neural patterns of meanings in each brain occur in trajectories of discrete steps, while the amplitude modulation in EEG wave is the mode of expressing meanings [1]. Zhou et al. have proposed some novel features for EEG signals to be used in brain–computer interface (BCI) system to classify left- and right-hand motor imagery [2]. The experimental results have shown that based on the proposed features, the classifiers using linear discriminant analysis, support vector machines, and neural network achieve better classification performance than the BCI-competition 2003 winner on the same data set in terms of the criteria of either mutual information or misclassification rate. Dressler et al. studied the anaesthetics on the brain and the level of sedation [3]. Lin et al. studied the change of human emotion during music listening through EEG signals [4].

The vast implications of using EEG data to analyze brain states include designing brain–computer interfaces (BCI) where users can operate on a machine via brain activities, and using brain state models in healthcare-related activities. Imagine a world where mere thinking about retrieving information will give you the results you are looking for; a world that no longer requires a keyboard, mouse, or traditional hardware input devices to interface with a computer; a place where you can instantly find out your health information in real time. Imagine a device that can instantly retrieve your body and mind condition and share this information with a medical expert that can then immediately analyze the data and make an appropriate health diagnosis. Instead of reacting to a condition that may already have caused irreparable health damage after the fact, there is a good chance that this information could be proactively provided and prevent deteriorating health conditions from occurring in the first place. Much of the capability and technology is available now to implement all these thoughts. We may not be able to know exactly what you are thinking, but we can gather brainwave data and make it available for analysis. We can control computers with mere thought! The methods may still be somewhat primitive and the technology in its infancy, but nevertheless, with only a few inexpensive off the shelve parts and a little ingenuity we can create a device that is capable of sensing body conditions and even read brainwaves.

As an example, we present a case study in transcendental meditation [5], a spiritual development technique, which was popularized by former Hindu ascetic Mharishi Mahesh Yogi and gained popularity in the west during the 1960s [6]. The concurrent brain states associated with transcendental meditation have been viewed as something outside of the world of physical measurement and objective evaluation by most scientific communities. Scientists now have the ability to measure and register electric potential of the human brain through the use of electroencephalographic technologies. One approach is to study finite differences within the minds of those practicing meditation, and those who do not. Such an endeavor is an avenue towards modeling a wide range of brain states [7]. The combination of electroencephalographic data with modeling methods in fields such as data mining and bioinformatics could be used to prove that subjects in a state of transcendental meditation are in a verifiable and observable state of mind that can be monitored and predicted [5]. Experiments found that cancer patients that practiced meditation experienced higher well-being levels, better cognitive function and lower levels of inflammation than a control group [8].

Challenges

The challenges in EEG-related studies include the design of the measuring tools and the methodologies in analyzing EEG data. Here we extricate a method to build an inexpensive headset to measure brainwaves. An EEG is a tool used to capture brainwave activity while it is performing a cognitive task. This allows the detection of the location and magnitude of brain activity involved in the various types of cognitive functions. EEGs allow the viewing and recording of the changes in brain activity during the time a task is performed. EEGs for this purpose have been around for many years, albeit only in medical research facilities and typically being very expensive. The intrigue is being able to inexpensively build an EGG with off-the-shelf parts and be able to perform the same type of brainwave research at home as sophisticated medical research facilities.

In addition, a platform for comprehensive EEG data storage and processing is desirable to promoting applications of using EEG tools in both physiological (e.g., clinical uses, sleep evaluation, fatigue detection, etc.) and psychological (cognitive sciences, BCI, etc.) scopes. Such a platform consists of EEG data collection devices (viz., EEG headset), communication channels (e.g., smart phones), a web server that provides a web interface for users to access stored EEG data and activate data analysis algorithms, and an online database for EEG data storage and processing.

The primary motivation behind this is to know what signals the brain produces does under certain situations and to know how these signals are consciously manipulated via controlled thoughts. Additionally it is desirable to know if there is a way to enhance studying and learning abilities and being able to retain more information. As an example, attempts have been made to study the tangible effects of meditation on human body and behavior, and investigate the possibilities of applying scientific methods to measure the effects. A direct benefit of this study will be to extend psychology to develop new methods for healing various mental diseases. This objective is feasible because meditation is efficient in training human self-control since its goal is having one's every whim under observation. Through this study, it is anticipated to start a campaign to establish "measurable" meditation methods, applying scientific methodology to religions, and eventually making religions "tangible".

Current Techniques

A cursory look into the topic revealed a wealth of information, much theoretical and limited to large government organizations and research facilities with huge budgets. For instance, the government has a program called the "Brain Research through Advancing Innovative Neurotechnologies™ (BRAIN)". The web site states the following: "The Brain Research through Advancing Innovative Neurotechnologies™ (BRAIN) Initiative is part of a new Presidential focus aimed at revolutionizing our understanding of the human brain. By accelerating the development and application of innovative technologies, researchers will be able to produce a revolutionary new dynamic picture of the brain that, for the first time, shows how individual cells and complex neural circuits interact in both time and space. Long desired by researchers seeking new ways to treat, cure, and even prevent brain disorders, this picture will fill major gaps in our current knowledge and provide unprecedented opportunities for exploring exactly how the brain enables the human body to record, process, utilize, store, and retrieve vast quantities of information, all at the speed of thought". The site even contains funding opportunities for companies and research facilities to participate and contribute to the program. Examples such as this can be found in abundance and what quickly becomes apparent is that there is a thirst for more knowledge about the human brain and how it works.

Very little information exists in the hobby and home space for EEG devices. Organizations such as OpenEEG and OpenBCI are available and facilitate the information sharing among hobbyists and attempt to inform the general public about the subject of gathering brainwave data. Companies like NeuroSky and Emotive sell headset EEG devices and provide software development kits (SDK) that include the tools necessary to gather brainwave data, but are limited to only reading brainwaves. In research perspectives, there is still space to gather more information, to have an enhanced data model, and see additional dependencies while the brain performs or reacts to specific tasks.

On the brain state modeling side, two types of research models have been used: statistical models and micro-models. Statistics models are built by applying statistical analysis to collected data from meditation practitioners, while micro-models try to catch physiological features of the brain state under examination. Current literatures show that both methods are used in the study of complementary and alternative medicine, which includes meditation as one of the methods. Loizzo et al., performed a 20-week contemplative self-healing program study, which showed that a contemplative self-healing program can be effective in significantly reducing distress and disability among the testers [9]. Habermann et al., on the other hand, performed a long-term (5–20 years) project to investigate the use of complementary and alternative medicine and its effects on the testers' health [10]. Comparisons across different groups of people are also found. For example, in a 6-week mindfulness-based stress-reduction program, subjects assigned to the program demonstrated significant improvements in psychological status and quality of life compared with usual care [11]. Another comparison is found where a group of Qigong practitioners were compared to a control group and positive indicators were found in the study [12].

A survey of the literature on cognitive impairment and cancer presented in [13] suggests that meditation may help improve cancer-related cognitive dysfunction and alleviate other cancer-related sequelae.

It is well understood that although statistical studies can provide evidence for the effectiveness of meditation, it fails to provide a systematic view of human's epistemology and psychology. This addresses the needs for micro-models that depict the inter-relationship between human's mind and physical body.

To accurately and objectively record moods when one is practicing meditation, we seek a solution which could objectively measure the effectiveness of meditation in real time. We start with a project that aims to create an application that takes EEG data and exposes it to various analytical techniques so the resultant brain states can be studied and predicted. We anticipate that, upon completion, this software can be used to produce important and dependable conclusions about a given subject's brain states and correlate that to an identified physical or psychological activity, and ultimately, we will be able to build a brain state model for meditation.

The concurrent brain states associated with transcendental meditation were viewed as something outside of the world of physical measurement and objective evaluation by most scientific communities. Due to the easily obtainable EEG headsets, recording EEG signals can be performed in a large scale. It is therefore possible to build a model for meditation brain state [14]. By applying data mining algorithms that quantify psychological states, we expect to analyze brain state associated with meditation to build models for meditation brain states [5]. Comparisons of the results obtained from different methods can be performed to fuse different models in order to have a deeper understanding of the central and peripheral nervous systems' role in attaining different levels of mediation. The practical significance of finding a meditation model includes establishment of the guidance for effective meditation exercises and a methodology for verifying the effectiveness of meditation methods. A scientific meditation model will advance studies in natural computer interface, identifying depression and mental illness, detect fatigue and boredom, and comprehending human emotion, etc. Any advances in these areas will have great social, economic, and technical significance.

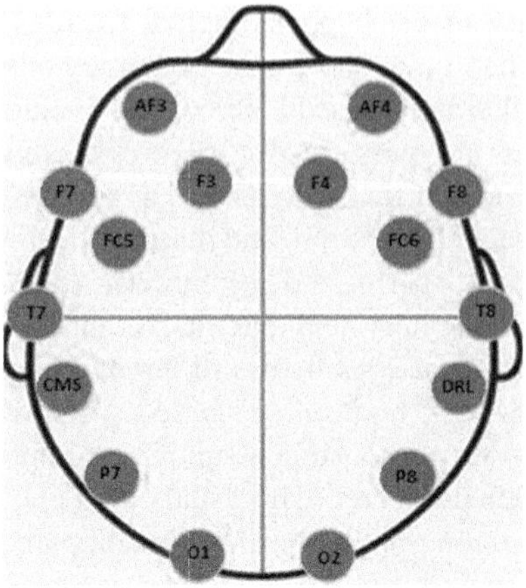

Fig. 1 Nodes AF3 through AF4 (counter clockwise)

We have conducted a trial collection of EEG data on a patient who performed MQ and a few other subjects performing tasks with different levels of brain activities, including the attempts of resting the brain. The Emotiv EPOC headsets we used can collect 14 channels of EEG signals. The locations of the contact points on the scalp, called nodes, are demonstrated in Fig. 1.

The data set analyzed so far comes from a study in which a candidate performed various tasks alternatively. The individual alternated between idle activity, reading news headlines, and participating in a mathematics exam limited to basic algebra every 60 s for 20 min. This data include 123,001 samples for all 20 min. With 129 samples per packet we have approximately 953 packets per node. We can assume that this leaves approximately 47 packets per node per minute. Then we apply linear regression to the generated scatter-plot, the positive or negative slope correlates to an increase or decrease in brain activity for the entire packet of said node. In the first packet of our data set, we notice correlations between two sets of four nodes ({AF3 F3 FC5 F7} and {AF4 F8 FC6 F4} respectively). By cross examining node position from Fig. 1 with the first packet of each node in Fig. 2, we can observe similar scatter-plots that are geometrically symmetric when referencing nodes regarding both left and right hemispheres of the frontal lobe. This tells us that {AF3 F3 FC5 F7} and {AF4 F8 FC6 F4} are sections of the brain that work together when the user is in an idle brain state. Our efforts are currently invested into recognizing repeatable patterns throughout the packets. Our developed signal processing algorithms will be used to determine repeatable characteristics in sets of packets that belong to each of the idle, news headline, and mathematics test states.

We also have developed a first version of an iOS application for iPad to analyze and visualize collected EEG data. This application can parse and visualize EEG data. It is also possible to extend this app to collect EEG data in real time, using a wireless EEG device. This will enable users to record, visualize, and analyze data in real time.

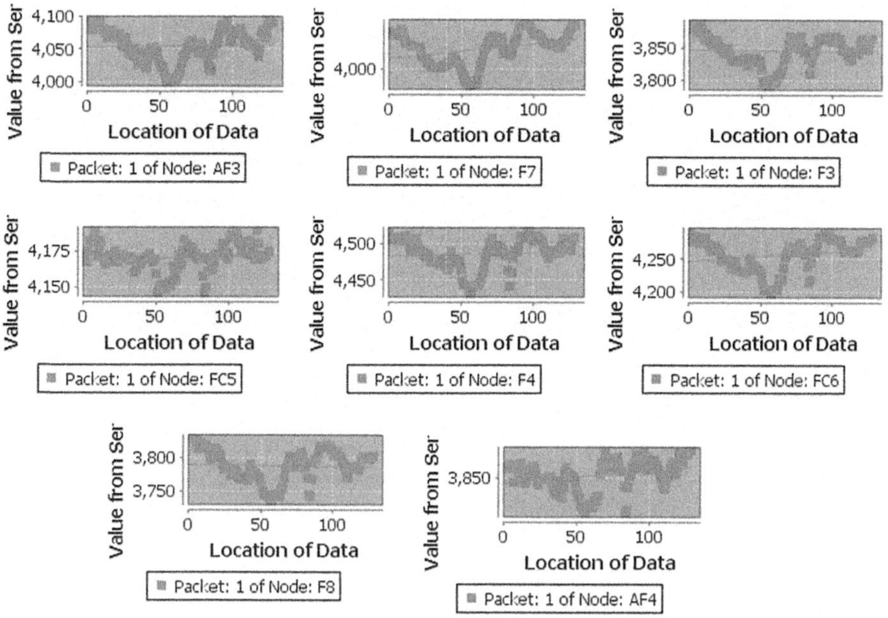

Fig. 2 Packet 1 of Nodes AF3, F7, F3, FC5, F4, FC6, F8, and AF4

Example One: Composition of EEG Headset

In this section we briefly describe how a simple EEG headset can be built using open source materials. The prototype multi-functional headset we built consists of an EEG sensor, a pulse sensor, a temperature sensor, a microprocessor, and a microprocessor blue tooth shield.

(1) EEG sensor, commercial product from NeuroSky. The NeuroSky technology was chosen for its dry sensors capabilities. This means that the sensor requires no special liquid chemicals while making contact with the skin to read brainwaves.
(2) Pulse Sensor, Open Source pulse sensor from pulsesensor.com. The pulse sensor is a current to voltage converter Op Amp circuit that uses a photodiode as current source. It has a low-pass Filter for output.
(3) Temperature sensor, commercial integrated circuit sensor, TMP36—Analog Temperature sensor from Adafruit. The TMP36 temperature sensor is a solid state device. Meaning it does not use mercury. Instead, it uses the fact that as temperature increases, the voltage across a diode increases at a known rate. By precisely amplifying the voltage change, it is easy to generate an analog signal that is directly proportional to temperature.
(4) Microprocessor: Arduino Mega 2560, Open Source.
(5) Microprocessor Blue Tooth Shield: Bluetooth Low Energy (BLE) Shield from redbear.com. Added to the Arduino for low-energy bluetooth communications with the iPhone.

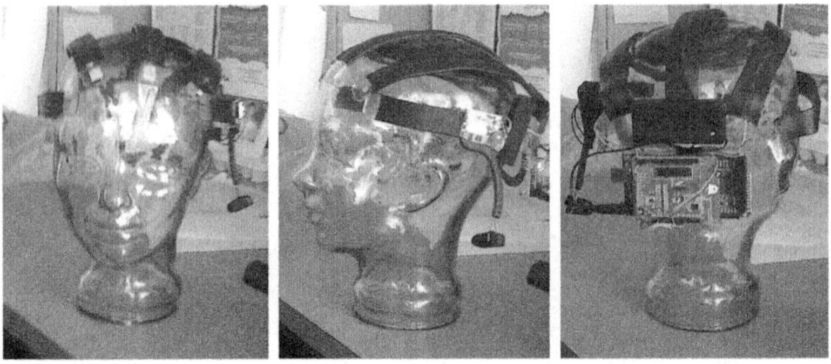

Fig. 3 Prototype multi-functional headset

The assembled headset is shown in Fig. 3, where the three sensors are mounted on the tips of three legs in the forehead, the microprocessor and the microprocessor bluetooth shield are mounted on the back, and the ear lobe is used as the base of the EEG sensor.

In order to test and validate that the headset is working properly and that all the sensors are functioning, a test environment had to be constructed. To simulate a real-world environment, a mobile Smart Phone application was developed on the Apple iPhone platform. This platform was chosen for ease of access to development tools and availability of software development kits (SDK) from all the hardware and chipset vendors. Both NeuroSky and Red Bear Labs included sample applications that were then easily transferred to a custom application using a simple view to display all the sensor values.

To show that the headset sensors are working a custom mobile application was developed to view the results. Sample screenshots of the application with the actual results are displayed in Fig. 4. The left snapshot shows the signal strengths of the

Fig. 4 iOS sensor headset application

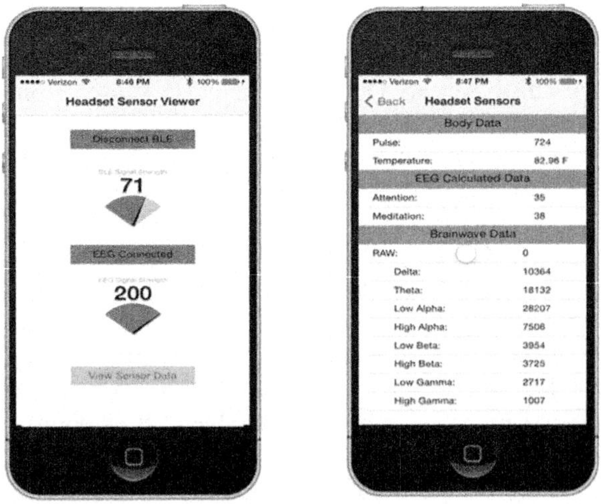

EEG sensor (the lower number) and the Arduino sensors (the upper number). Readings of the three sensors are shown on the right snapshot. Since we use the NeuroSky EEG sensor, the EEG signals are filtered into signals of different frequency intervals.

Example Two: Analysis of Meditation State

The brain emits electrical signals that are caused by neurons firing in the brain. The patterns and frequencies of these electrical signals can be measured by placing a sensor on the scalp. For example, the EEG sensor by NeuroSky is able to measure the analog electrical signals commonly referred to as brainwaves and process them into digital signals to make the measurements available for further analysis. Table 1 lists the most commonly recognized frequencies that are generated by different types of brain activity.

Emotions play an essential role in many aspects of our daily lives, including decision-making, perception, learning, rational thinking and actions. To detect the emotion of a person, the first approach is based on text, speech, facial expression, and gesture. This approach, needless to say, is not reliable to detect emotion, especially when people want to conceal their feelings. Some emotions can occur without corresponding facial emotional expressions, emotional voice changes, or body movements. On the contrary, such displays could be faked easily. Using multi-modality approach can overcome this shortcoming to limited extent.

The new approach is through affective computing, which employs EEG signals recorded when users perform some brain activities and apply analytical algorithms to EEG data to detect the emotion. This approach is based on the fact that brain activities have direct information about emotion and EEG signals can be measured at any moment and are not dependent on other activities of the user such as speaking or generating a facial expression. Different recognition techniques can be used in different situations to maximize recognition rates.

Table 1 Brainwave frequencies

Brainwave type	Frequency range (Hz)	Mental states and conditions
Delta	0.1–3	Deep, dreamless sleep, non-REM sleep, unconscious
Theta	4–7	Intuitive, creative, recall, fantasy, imaginary, dream
Alpha	8–12	Relaxed, but not drowsy, tranquil, conscious
Low Beta	12–15	Formerly SMR, relaxed yet focused, integrated
Midrange Beta	16–20	Thinking, aware of self and surroundings
High Beta	21–30	Alertness, agitation

An Empirical Study

We measured an experienced meditator's brainwaves while meditating and compared them to several other states including idle and talking. We found prominent differences between the experienced meditator's brainwaves and those of other states. The experienced meditator's brainwaves clearly displayed a stable state most of the time, as shown in Fig. 5a. However, during certain times after the initial meditation stage, extraordinary high waves were observed, as shown in Fig. 5b.

Figure 6 shows the brainwaves of idle, talking, and meditating from an inexperienced meditator. We can clearly see that the irregularities of these states are higher than the experienced meditator's state, especially the idle and the talking states. The inexperienced meditator showed some similarity to the state shown in Fig. 5a but it did not show the features in Fig. 5b. This initial study indicates that trained meditators can demonstrate regularity during meditation practice.

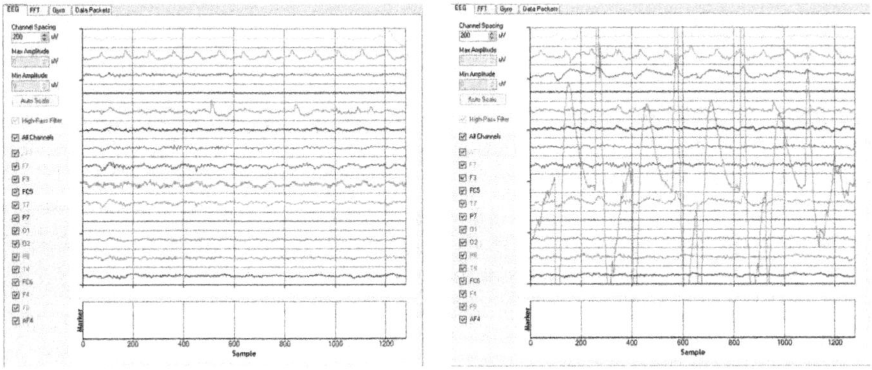

(a) Meditating 1 **(b)** Meditating 2

Fig. 5 An experienced meditator's brain waves

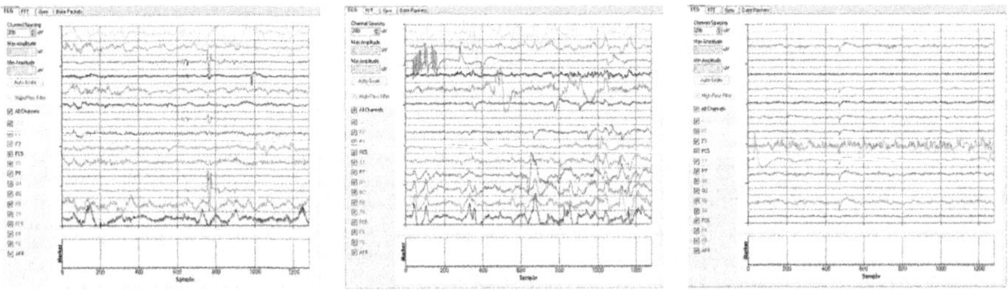

(a) Idle **(b)** Talking **(c)** Inexperienced Meditation

Fig. 6 Brain waves of other states

A Platform for Cumulative Brain State Modeling

We are investigating an automatic EEG-based emotion recognition system that can record the EEG signals from users and measure their emotions. The EEG data are filtered to get separate frequency bands to train emotion classifiers with the four well-known classification techniques that are SVMs, Naïve Bayes, *k*NN and AdaBoost.M1. Figure 7 shows the typical flowchart of data processing.

Table 2 shows the brain state recognition rate of different algorithms and Table 3 shows the band-wise recognition rate of the AdaBoost.M1 algorithm.

As depicted above, using EEG data, cognitive psychologists can visualize and observe correlations between different active brain states. It is desirable to create an application that takes EEG data and exposes it to various analytical techniques so the resultant brain states can be studied and predicted. We present the design and implementation of a system that integrate onsite EEG data collection, analysis, web-based EEG data storage and modeling tools, and user feedback through mobile communication devices. Architecture of the system is shown in Fig. 8.

The web server provides a user interface that allows users to view EEG data in the database and run R program to perform data analysis. Figure 9 shows that data

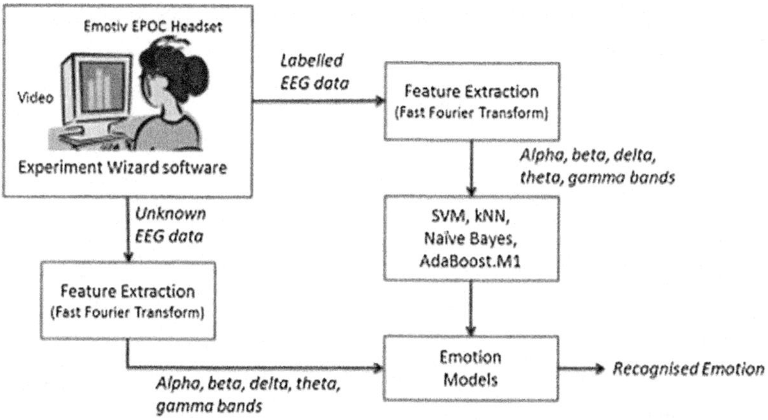

Fig. 7 EEG data analysis flowchart

Table 2 Brain state recognition rates

	SVM (%)	*k*NN (%)	Naïve Bayes (%)	AdaBoost.M1 (%)
Emotion recognition rate	89.25	83.35	66	92.8

Table 3 Recognition rates of AdaBoost.M1

	Delta (%)	Theta (%)	Alpha (%)	Beta (%)	All (%)
Emotion recognition rate	69.95	68.4	75.5	89.7	92.8

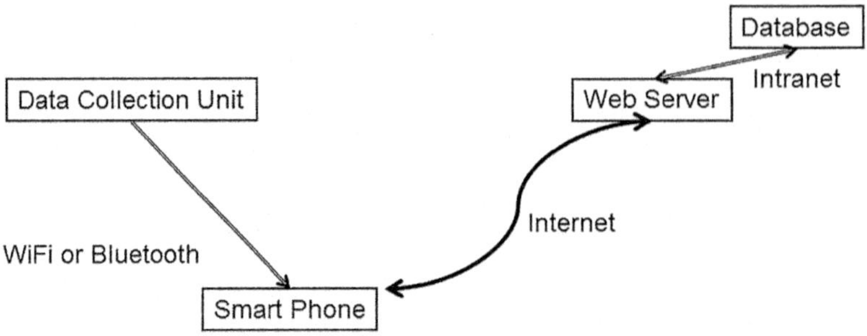

Fig. 8 EEG data analysis system architecture

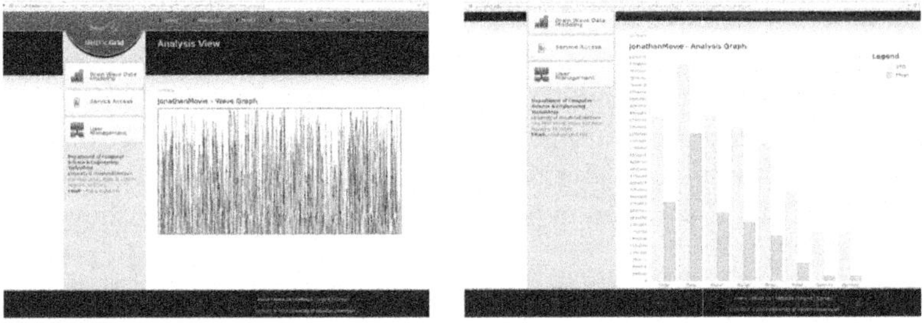

(a) EEG Wave forms **(b)** EEG Histogram

Fig. 9 EEG data Rendering on web interface

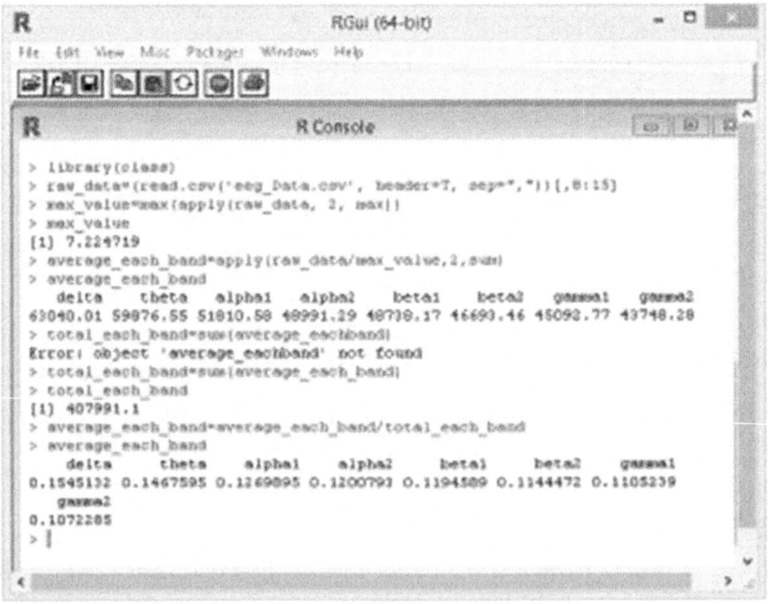

Fig. 10 Running R to Analyze EEG Data

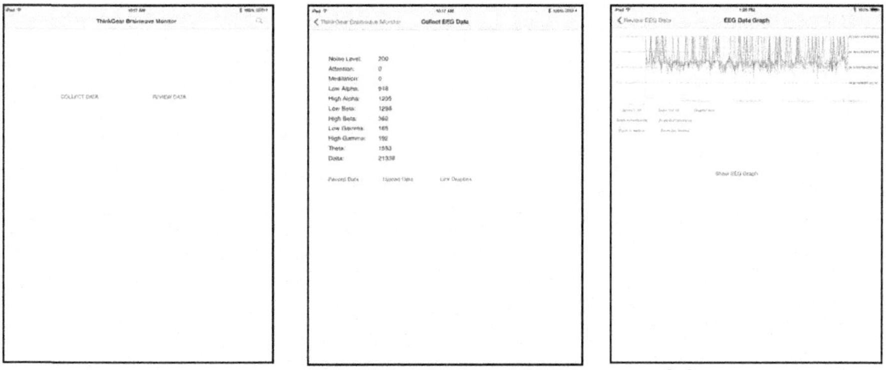

(a) Two functions (b) Text mode display (c) Wave form

Fig. 11 iPhone interface to the web server

are rendered in wave form mode and statistical mode, respectively. Figure 10 shows that R is invoked to perform data analysis tasks in interactive mode.

iPhone users can also use the web interface to connect to the database to view data. Figure 11 shows the two functions, viz., "collect data" and "view data", that a user can choose on the iPhone app (Fig. 11a). The user can display data in text mode by viewing individual data frames (Fig. 11b), or display the wave form of recorded data in certain time period (Fig. 11c).

Conclusion

We presented some information depicting the current status of EEG research with projected applications in the areas of health care. We hope this information provides the reader a bird's eye view of the potentials of EEG data analysis.

In addition, we described a method of quick prototyping an EEG headset, in a cost-effective way and with state-of-the-art technologies. This headset is a good starting foundation for anybody interested in researching body data via sensors. The Arduino components can be extended and exchanged to any desired configuration. It is only up to the imagination of the builder to decide what is possible and where to take the project next.

We used meditation research to reach out to the high-end applications of EEG data analysis in understanding human brain states and assisting in promoting human health care. Some devotees to the practices of transcendental meditation have shown the ability to control these brain states. We want to numerically prove or disprove this assumption; the analysis of these states could be the initial step in a process to first predict and later allow individuals to control these states.

References

1. W.J. Freeman, A neurobiological interpretation of semiotics: meaning, representation, and information. Inf. Sci. **124**(2000), 93–102 (2000)
2. S.-M. Zhou, J.Q. Gan, F. Sepulveda, Classifying mental tasks based on features of higher-order statistics from EEG signals in brain-computer interface. Inf. Sci. **178**(6), 1629–1640 (2008)
3. O. Dressler, G. Schneider, G. Stockmanns, E.F. Kochs, Awareness and the EEG power spectrum: analysis of frequencies. Br. J. Anaesth. **93**(6), 806–809 (2004)
4. Y.-P. Lin, C.-H. Wang, T.-P. Jung, T.-L. Wu, S.-K. Jeng, J.-R. Duann, J.-H. Chen, EEG-based emotion recognition in music listening. IEEE Trans. Biomed. Eng. **57**(7), 1798–1806 (2010)
5. R. Davidson, J. Kabat-Zinn, J. Schumacher et al., Alterations in brain and immune function produced by mindfulness meditation. Psychosom. Med. **65**(4), 564–570 (2003)
6. B.K. Holzel, J. Carmody, M. Vangel, et al., Mindfulness practice leads to increases in regional brain gray matter density. Psychiatry Res. Neuroimaging. 2 022 **191**(1), 36–43 (2010)
7. H. Lin, Measurable Meditation, in *Proceedings of the International Symposium on Science 2.0 and Expansion of Science (S2ES 2010), the 14th World Multiconference on Systemics, Cybernetics and Informatics (WMSCI 2010)*, Orlando, Florida, June 29–July 2, 2010, pp. 56–61
8. B. Oh, P. Butow, B. Mullan et al., Impact of medical Qigong on quality of life, fatigue, mood and inflammation in cancer patients: a randomized controlled trial. Ann. Oncol. **3**, 608–614 (2009)
9. J.J. Loizzo, J.C. Peterson, M.E. Charlson, E.J. Wolf, M. Altemus, W.M. Briggs, L.T. Vahdat, T.A. Caputo, The effect of a contemplative self-healing program on quality of life in women with breast and gynecologic cancers. Altern. Ther. Health Med. **16**(3), 30–37 (2010)
10. T.M. Habermann, C.A. Thompson, B.R. LaPlant, B.A. Bauer, C.A. Janney, M.M. Clark, T.A. Rummans, M.J. Maurer, J.A. Sloan, S.M. Geyer, J.R. Cerhan, Complementary and alternative medicine use among long-term lymphoma survivors: a pilot study. Am. J. Hematol. **84**(12), 795–798 (2009)
11. C.A. Lengacher, V. Johnson-Mallard, J. Post-White, M.S. Moscoso, P.B. Jacobsen, T.W. Klein, R.H. Widen, S.G. Fitzgerald, M.M. Shelton, M. Barta, M. Goodman, C.E. Cox, K.E. Kip, Randomized controlled trial of mindfulness-based stress reduction (MBSR) for survivors of breast cancer. Psychology **18**(12), 1261–1272 (2009)
12. B. Oh, P. Butow, B. Mullan, S. Clarke, Medical Qigong for cancer patients: pilot study of impact on quality of life, side effects of treatment and inflammation. Am. J. Chin. Med. **36**(3), 459–472 (2008)
13. K.A. Biegler, M.A. Chaoul, L. Cohen, Cancer, cognitive impairment, and meditation. Acta Oncol. **48**(1), 18–26 (2009)
14. H. Lin, Measurable Meditation, in *Proceedings of the International Symposium on Science 2.0 and Expansion of Science (S2ES 2010)*, Orlando, Florida, USA, pp. 56–61 (2010)

Big Health Data Mining

Chao Zhang, Shunfu Xu and Dong Xu

Abstract With the improvement of infrastructures and techniques, "Big Data" provides great opportunities to health informatics, but at the same time raises unparalleled challenges to data scientists. As an interdisciplinary field, the health data are not limited to electronic health record (EHR), as more and more molecular-level data are used for disease diagnosis and prognosis in clinics. Effectively integrating and mining these data holds an indispensable key to translate theoretical models into clinical applications in precision medicine. In this chapter, we briefly demonstrate different data levels involved in health informatics. Then we introduce some general data mining approaches applied to different levels of health data. Finally, a case study is illustrated as an example for applying computational methods on mining long-term EHR data in epidemiological studies.

Introduction

"Big Data" is a new buzzword without a consensus on its definition, and the properties of Big Data have changed from 3Vs, Volume, Velocity, and Variety to 5Vs adding Value and Veracity [13]. No matter how much the definition changes,

C. Zhang (✉)
Institute for Computational Biomedicine, Weill Cornell Medicine,
New York, NY 10021, USA
e-mail: chz2009@med.cornell.edu

C. Zhang
Division of Hematology and Medical Oncology, Department of Medicine,
Weill Cornell Medicine, New York, NY 10021, USA

S. Xu
Department of Gastroenterology, The First Affiliated Hospital of Nanjing Medical University,
Nanjing 210029, Jiangsu, China

D. Xu
Department of Computer Science and C.S. Bond Life Sciences Center,
University of Missouri, Columbia, MO 65211, USA

D. Xu et al. (eds.), *Health Informatics Data Analysis*, Health Information Science,
DOI 10.1007/978-3-319-44981-4_11

no one can deny that health informatics has entered the Big Data era. Previous studies have detailed that the data in health informatics exhibit all the properties of 5Vs. Big Data not only leads to a major opportunity to create profit for business intelligence in many industries, such as retail outlets and social media, but also creates a very challenging problem in the data mining field. With the broader usage of electronic health records (EHR) and the explosion of other levels of data in the past two decades, providing health information with easy access and maintenance has become critical for both patients and physicians. Like other industries, the secondary usage of rich health data has provided incentives to significantly accelerate new drug development and clinical trials for pharmaceutical companies [37]. In addition to its potential marketing value, large-scale health data are also a valuable asset for research, although there are some challenges that must be overcome [27]. Mining health data with suitable platforms or algorithms may get the optimal clinical treatment thereby building connections between phenotypic and genomic information effectively [15, 38]. These connections are supplemented by Big Data provision of long-term records with unprecedented coverage, which can extend clinical epidemiological studies from a city to a state, and even to other countries [1, 34]. With the rapid improvement of biotechnologies, health data are not limited to EHR, as more and more molecular-level data are used for disease diagnosis and prognosis in clinics. Hence, integrating and mining the information from different sources or levels are urgent challenges in health informatics.

In this chapter, we briefly introduce the different data levels involved in health informatics. We then list a number of algorithms used in health data analysis to provide some introduction on health data mining. Finally, instead of demonstrating a specific algorithm or platform for investigating the relation between health information and a particular phenotype, we illustrate a case study for applying statistical learning methods on long-term EHR data in epidemiological studies. In particular, 127,173 cases across eight years were accessed to evaluate the risk factors of intestinal metaplasia (IM) for two provinces of southeastern China. IM is closely related to the occurrence of intestinal-type gastric carcinoma (GC), which is the third leading cause of cancer mortality in China [4]. Our research focused on a comprehensive assessment of the status of IM, which helped examine the risk factors associated with IM.

Data Level

Traditional health informatics focused on the mining or studies of EHR data [20], but nowadays it has already spanned to many data types using methods from multiple disciplines. The borderlines between bioinformatics, biostatistics, medical informatics, and health informatics are not clear anymore. Different levels of data carrying different characteristics are utilized by different organizations and have been analyzed by different mining methods. Here, we briefly introduce the data levels involved in health informatics.

Molecular Level

During the past decade, the largest "deep" molecular-level datasets are generated by The Encyclopedia of DNA Elements (ENCODE) [11] and The Cancer Genome Atlas projects [6, 7]. The ENCODE project has generated more than 2600 genomic datasets from multiple platforms, such as ChIP-seq, RNA-seq, ChIA-PET, CAGE, high-C, and so on [22]. Although it is just a start, TCGA has already collected more than 8200 tumor samples with measurements of somatic mutations, copy number variations, mRNA and miRNA expression, and protein expression. With the help of these data, the role of aneuploidy (abnormal number of chromosomes) in cancer development, which was a 100-year-old puzzle in cancer research raised by German biologist Theodor Boveri in 1914, was finally answered by geneticist Stephen Elledge [42].

Beside using the above data to solve some basic biological problems, these genomics data and other '-omics' data have been integrated with clinical data to provide a platform to improve the treatment in practice, and to discover the complete drug interactions and outcomes in particular populations which could be missed by clinical trials. For example, cBioPortal [9], an open-access resource for interactive exploration of multidimensional cancer genomics data sets, contains more than 18,000 tumor samples from 80 cancer studies. It transforms terabytes of molecular data from cancer tissues and cell lines into readily understandable genetic, epigenetic, gene expression, and proteomic events, and it also provides an intuitive Web interface for querying samples according to different features from genetic characteristics to clinical outcomes. With the graphical summaries of gene-level data, it can lower the barriers between biomedical research and clinical application, and it also makes the data easily accessible to researchers and clinicians without informatics expertise.

Tissue Level

Unlike the molecular-level data which details the observations of a small group of specific cells [35], tissue-level data presents the global behavior of multicellular organisms. Typically, most tissue-level data are imaging data. With the improvement of imaging technology, clinicians and research are not only able to examine the 2D data of pathology slides and 3D data of patient organisms, but also the 4D data of dynamically changing mammalian organs [3]. In clinic, high-resolution medical imaging instruments are widely used for disease diagnosis and monitoring. Large volumes of imaging data are generated by pathology and radiology departments, and most of them are from MRIs, Computed Tomography, and X-rays.

Real-time imaging provides more informative data. New analyses of these data can help map the borders and growth of tumors, and then deliver precise doses of chemotherapy or radiation within those borders. Aside from the common challenges of big data, such as storage and transmission, accurately and effectively processing imaging data will be helpful to shorten the lag time of medical diagnosis, although it is very challenging in medical practice. Many different learning algorithms have been employed to address the different problems, such as medical image analysis, computer-aided diagnosis, image reconstruction, and image retrieval.

Patient/Person Level

"Patient data come of age" has first been mentioned in the May 2003 issue of Pharmaceutical Executive (http://www.pharmexec.com/patient-data-come-age). At the beginning, data were derived from 1.6 billion prescriptions in the United States annually. The data were studied by pharmaceutical companies to track the prescription activity at the patient level to adjust marketing strategy. With the wide usage of EHR, physicians can easily access any patient's medical history including a history of his or her medications to decide therapy changes or dosing changes. In some extreme situations, a large volume of real-time patient-level data is generated, such as the data generated during ICU. These data not only require accurate real-time processing and analyzing to give treatment advice to the physicians, but also provide great value for long-term studies, such as the prediction of patient mortality rate after ICU discharge in 5 years.

Unlike EHR data generated by hospitals, with the improvement of wearable technology, an incredible amount of person-level data is coming up and generated by users themselves. According to ABI Research forecasts, 28 million smart watches would be shipped in 2015 (https://www.abiresearch.com/blogs/apple-watch-forecasts/). Other types of wearable devices were showcased, such as Smart watches, Smartbands, Smart Jewelry, glasses, and ear buds. These revolutionary devices provide unprecedentedly real-time monitoring to not only patients, but also healthy people. Precision processing of these data can provide most accurate and comprehensive health and fitness guidance to users, and this information can also be integrated with EHR data to help physicians to diagnose diseases and take preventive measures for potential health issues.

Population Level

Traditionally, population-level data in health informatics are gathered from a hospital or clinical researchers, and has been used for answering both clinical questions and epidemic-scale questions. With the improvement of EHR and data mining techniques, these data accumulated for years could be used to give some old questions a more accurate answer, such as the relationship between risk of gastrointestinal bleeding and long-term use of aspirin [25]. They can also lead to some new discoveries, such as reducing risk of colorectal cancer with aspirin [10].

Besides the increasing EHR data, the explosion of other types of Big Data also provides additional information for epidemiological studies, such as the data from Twitter, Facebook, or Google search data. By integrating human mobility data, Lemey et al. [30] created a better model to predict the global transmission dynamic of the human influenza virus H3N2.

Currently, population-level data are not limited to EHR data anymore and covers all large-scale studies of all data levels. Integrating and mining different levels of data is a most challenging task in health informatics, and it also provides a great opportunity to enhance the patient care system and improve public health. CancerLinQ [45], a platform developed by The American Society of Clinical Oncology (ASCO), keeps gathering cancer patient records including genomic data, diagnoses, and notes on treatment, and measures how well patients respond to therapy. Through learning these big data, the system will keep improving and finally provide the optimal cancer care clinical guidelines.

Heterogeneity of Big Health Data

Heterogeneity is immanent in big data. On one hand, heterogeneity refers to the variant data sources. In health informatics, heterogeneous data could be either from multiple medical centers or from multiple data levels. Integrating heterogeneous data from different sources is the first and very important step of data analysis [29]. Using data from multiple sources might lead to better results than only using one single data. Especially in healthcare system, disease diagnosis for each patient always relies on multiple data sources, so the accumulated big health data are also from heterogeneous sources. These data always have different structures, and some of them could be incomplete or with errors. Carefully processing and integrating them remain challenging. On the other hand, big health data might represent the heterogeneous information from different subpopulations, so failing to take heterogeneity into account can easily derail the discoveries from these data. Shah and coworkers collected 397 heart failure with preserved ejection fraction cases with 46 continuous phenotypic variables, and then they used computational method to analyze the data to reveal three mutually exclusive subgroups [43]. Similarly, Wang and colleagues integrated multiple data for 1500 patients to discover three novel subgroups of gastric cancer [48].

Techniques

In terms of methodologies used in health data mining, almost all statistical or learning-based methods have been employed to solve many different problems in this field.

Statistic-based methods, especially most regression methods, were widely used to conquer the most common problem wherein the "parameters *P* is much larger than the number of observations *N*" (e.g., the number of genomic variations is much larger than biomedical sample size) in genome-wide association studies (GWAS) [8]. GWAS exploits the significance of associations between small DNA variations (Single Nucleotide Polymorphism, SNP) and phenotype, by assuming "common disease, common variant". In order to accurately and effectively explore the relationships between millions of SNPs and complex traits/disorders, researchers extended the methods from simple linear regression to many advanced models, such as Bayesian linear regression [33], linear mixed models [32], penalized multiple regression [24], and so on. Those studies have suggested previously unknown markers or pathways associated with some diseases, such as IL23R with Crohn's disease [17] and FGFR2 with breast cancer [26]. Another widely used model, the generalized linear model, has been adapted by many software packages for differential analyses on gene expression or DNA methylation. For example, an unexpected role of miR-7 in cortical growth through its interactions with p53 pathway genes was discovered through a differential gene expression analysis between the wild-type and miR-7 silenced mouse models [40].

Supervised learning (classification) model uses labeled data (training data) to produce an inferred function, and then applies it to classify the new samples (unlabeled test data). We can easily find examples for utilizing popular supervised learning models, such as Support Vector Machines (SVM), Random Forest, Neural Networks, and so on. Zhang et al. [50] formulated a residues-based classifier to access the gastric cancer risk by the *Helicobacter pylori* (*H. pylori*) marker gene. The authors extracted the features from the sequences of the most important virulence factor CagA of *H. pylori* and then trained an SVM model with selected key residues. They tested their model with leave-one-out cross validation, and the result indicated the relationship between pathogen sequence marker variation and cancer risk. In another study, Michaelson et al. [36] analyzed whole genome sequencing data from 10 pairs of monozygotic twins concordant for autism spectrum disorders. They extracted de novo mutations (DNMs) in affected individuals, and then trained a Random Forest classifier to evaluate the importance of DNMs contributed to autism. Their findings suggest that regional hypermutation is a significant factor involved in autism. Sometimes multiple models have been employed in a single study to get better performance. For example, Han et al. [23] generated pseudogene expression profiles in 2808 patient samples from TCGA dataset, and then they applied three supervised learning methods to predict the cancer subtypes using pseudogenes with the most variable expression. They found a significant number of pseudogenes with different expression levels in different cancer subtypes, which can be used to classify the histological cancer subtypes.

Unsupervised learning (clustering) model attempts to find the underlying structure in unlabeled data, and it is the best model to discover the molecular types of some not well-studied diseases. For example, principal components analysis (PCA) is an unsupervised approach to reduce the dimensionality of data while identifying hidden features with the most signals, and the first principal component has the largest possible variance. PCA analysis has been performed in most large-scale studies, such as breast cancer studies from TCGA [5]. Besides traditional unsupervised learning algorithms, such as PCA and K-means, some complex methods have been developed for integrative clustering samples with different levels of data. iCluster, a joint multivariate regression model, was originally proposed to cluster lung cancer samples from TCGA using both copy number and gene expression data, and then identify the subtypes characterized by concordant genes from two levels of molecular data. In a recently published gastric adenocarcinoma [7] study from TCGA, DNA methylation data were also analyzed with copy number, mRNA expression, miRNA expression, and protein expression by iCluster; moreover, gastric cancer cases have been divided into four subtypes: Epstein–Barr virus positive, microsatellite instability, genomically stable, and chromosomal instability.

Other methods, such as the algorithms from information theory and graph theory, are also widely used to process molecular data. Shannon's entropy and Boltzmann's entropy were employed by different research groups multiple times to analyze DNA methylation pattern in diseases. Xie et al. [49] calculated the Shannon's entropy of the methylation pattern of every four adjacent CpG sites to evaluate the epigenetic heterogeneity of diseases and find the regions more accessible for DNA methyltransferases. Li et al. [31] quantified the changes using the Boltzmann's entropy difference of every four adjacent CpG sites between diagnosis and relapse samples from leukemia patients, and according to the differences they demonstrated that the global clonality shift might drive the leukemia relapse. A Bayesian network model has been used to identify cancer driver genes in the research work of Akavia et al. [2]. They integrated chromosomal copy number and gene expression data for detecting aberrations that promote cancer progression, and they further confirmed several known drivers in melanoma samples.

A Showcase

Here we present a study wherein 127,173 upper endoscopies were performed from 2004 to 2011 as an example of mining long-term EHR data. Our research focused on completing a comprehensive assessment of the status of intestinal metaplasia (IM) and evaluating the risk factors of IM for two provinces of southeastern China.

Background

The major symptom of gastric IM involves gastric epithelium and gastric glands that transform into intestinal epithelium and intestinal glands, respectively, under pathological conditions. The most recognized theory by Correa [16] on the correlation between IM and GC focuses on the occurrence of intestinal-type GC where an inflammatory process in the gastric antrum is considered to be the cause of the initial lesion, which may progress toward chronic atrophic gastritis (AG), IM, dysplasia, and finally the invasive carcinoma. Thus, IM is the precancerous lesions of GC. Although in recent years the occurrence of GC has decreased, China remains a high-risk area. In 2008, the age standardized mortality rate for gastric cancer in China was 0.486‰, three times higher than the average global rate. The central region of Jiangsu Province in particular is a high-risk area for GC with a standardized incidence rate of 1.45‰ in Yangzhong City of the Zhenjiang municipal area, which also has a GC mortality rate of 0.79‰. Therefore, in this region, the prevention of GC is still very challenging.

Data and Methods

Records of gastroscopic results in the Digestive Endoscopy Center of the First Affiliated Hospital of Nanjing Medical University from 2004 to 2011 were retrieved from their Endoscopy Information System, including the patients' age, gender, images of endoscopic examinations, endoscopic and pathological findings, and the results of rapid urease tests. All data were collected in accordance with the institutional ethical and clinical guidelines. Personal information will not be disclosed in this chapter. The following items were assessed according to the updated Sydney classification system: chronic and acute inflammation, gastric glandular atrophy, and IM. All items were scored from 0 (absent) to 1 (mild), 2 (moderate), or 3 (marked) [46]. Gastric appearance and histological results were both used for diagnosis of gastric ulcer (GU), duodenal ulcer (DU), bile reflux, gastritis, IM, GC, and so on. Multiple analysis methods have been applied to the data for evaluating the relations between different factors. Odds ratio (OR), chi-square test, and t-test were carried out to examine correlations between IM status and *H. pylori* infection, AG, dysplasia, age, gender, peptic ulcer, bile reflux, chronic inflammatory severity, degree of acute inflammation, and lymphoid follicle number. The Cochran–Armitage trend test of IM was also carried out. A linear regression analysis was applied to geographical information. All p-values calculated were two-tailed and at a significance level of 0.01.

Previous Reports About the Basic Risk Factors of IM

Risk factors of IM have been the subject of several studies. De Vries et al. [12] reported that risk factors might increase the incidence of IM, including age, smoking, obesity, drinking, *H. pylori* infection, and bile reflux. Den Hoed et al. [14] found no differences in patients with IM and normal gastric mucosa with respect to gender, but subjects with IM were significantly ($P < 0.001$) older than subjects with either normal gastric mucosa or non-atrophic gastritis. Peleteiro et al. [39] showed that in smokers infected with high-virulence *H. pylori* strains, the risk of IM was further increased. Felley et al. [19] showed obesity, which was BMI >25 kg/m^2 in males and BMI >27 kg/m^2 in females, was also one of the risk factors of IM.

Age

In our study, we grouped the patients into six age groups, <20, 20–29, 30–39, 40–49, 50–59, and $\geq$60, respectively. As shown in Table 1, the impact of age on IM also reflected that the older the patient, the higher the incidence rate of IM and the higher the level of IM. The conclusion was consistent with findings of previous studies [12, 14]. A Cochran–Armitage test was performed between any two age groups to measure the trends between the IM development and different age groups. With increasing age, not only did the incident rate increase, but also the severity of IM (Fig. 1). It was obvious that IM was not just a common occurrence associated with aging. This may be due to a variety of known and unknown risk factors, especially for *H. pylori* infection, the roles of which accumulate with increasing age, leading to the appearance and aggravation of IM.

Gender

Gender was reported to be a risk factor for gastric cancer by previous studies [12, 41], but some previous studies also implied that there was no significant relationship between gender and IM, AG, or dysplasia [14]. Our large sample study revealed that gender might be an independent risk factor for IM. In our study, the IM incidence rate in the male population was 0.62% higher than in the female population, and the chi-square test between incidence rates of IM in males and in females was significant (OR, 1.04; 95% CI: 1.00–1.07; P = 0.03).

Other Diseases

For the correlations between IM and other gastric diseases, the literature reported that IM occurring on the basis of AG was generally recognized [28], but the relationship between IM and bile reflux had rarely been reported. Tsukui et al. [47] reported DU disease reduced the risk of contracting both IM and AG conditions, but not GU. Dysplasia is a mucosa lesion, which occurs on the basis of IM, and could progress to GC [28].

In our study, AG showed an extremely high positive correlation with IM incidence. Comparing AG with non-AG cases, the IM incidence rate was 95.70% versus 18.88%. Except for DU with significant negative correlation, the rest of the

Table 1 Statistics of risk factors based on IM scores

Characteristics	Absent (0)		Mild (1)		Moderate (2)		Severe (3)		*P* value
	Num.	%	Num.	%	Num.	%	Num.	%	
Age									
<20	1442	2.17	51	0.34	5	0.14	0	0	
20–30	6189	9.33	397	2.63	46	1.33	1	0.42	<0.01
30–40	13,580	20.46	1757	11.66	213	6.18	4	1.69	<0.01
40–50	15,166	22.85	3268	21.68	601	17.43	36	15.19	<0.01
50–60	15,396	23.20	4548	30.17	1075	31.17	82	34.60	<0.01
>60	14,592	21.99	5052	33.52	1509	43.75	114	48.10	<0.01
Gender									
Male	34,921	52.62	7903	52.43	1993	57.78	145	61.18	
Female	31,444	47.38	7170	47.57	1456	42.22	92	38.82	<0.01
Gastric ulcer									
Ulcer	3145	5.55	1313	9.73	351	11.16	31	14.42	
Gastritis	53,508	94.45	12,175	90.27	2794	88.84	184	85.58	<0.01
Duodenal ulcer									
Ulcer	4630	7.96	928	7.11	111	3.82	6	3.16	
Gastritis	53,508	92.04	12,175	92.89	2794	96.18	184	96.84	<0.01
Bile reflux									
Reflux	4400	6.63	1036	6.87	279	8.09	20	8.44	
Non-reflux	61,965	93.37	14,037	93.13	3170	91.91	217	91.56	<0.01
H. pylori									
Positive	20,428	33.37	5661	39.98	1151	35.59	67	31.60	
Negative	40,797	66.63	8497	60.02	2083	64.41	145	68.40	<0.01
Atrophic gastritis									
None	66,234	99.80	13,536	91.72	1813	54.18	70	30.04	
Mild	99	0.15	1066	7.22	832	24.87	47	20.17	<0.01
Moderate	26	0.04	145	0.98	688	20.56	89	38.20	<0.01
Severe	6	0.01	11	0.08	13	0.39	27	11.59	<0.01
Dysplasia									
None	63,020	94.99	12,843	87.03	2694	80.54	166	71.24	
Mild	2069	3.12	1663	11.27	574	17.16	58	24.89	<0.01
Severe	1253	1.89	251	1.70	77	2.30	9	3.86	0.19
Gastric cancer									
Cancer	3722	5.61	587	3.89	177	5.13	12	5.06	
Noncancer	62,643	94.39	14,486	96.11	3272	94.87	225	94.94	<0.01

factors were all positively correlated with IM incidence ($P < 0.01$). According to the OR, they had been ranked as follows: dysplasia, GU, *H. pylori* infection, and bile reflux.

A Cochran–Armitage test was performed between any disease and contrast groups to measure the trends between different diseases and IM grades (Table 1).

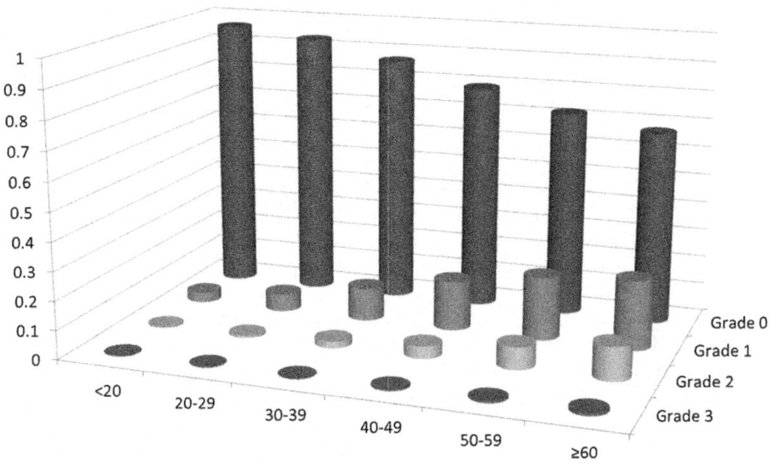

Fig. 1 Trends of IM development in different age groups

The relationships of IM grade to bile reflux, GU, DU, dysplasia grade, and AG grade were described. IM grade increased with GU ($P < 0.01$), bile reflux ($P < 0.01$), and the rising of AG grade ($P < 0.01$). The trend of the IM grade correlated with low grade dysplasia ($P < 0.01$), but not high-grade dysplasia ($P = 0.19$). There was a lower IM grade in patients with DU than in patients without it ($P < 0.01$).

Several previous studies have reported the correlation between GC and IM [44]. Since a biopsy was not performed on relatively normal parts of the stomach, our results showed a lower incidence of IM in GC patients. But for endoscopic diagnosed GC which also had non-tumor pathological data, the IM incidence was high. This also showed a significant correlation between GC and IM [21].

Other Potential Factors

In addition to the regular risk factors discussed up to this point, we also evaluated relationships between IM occurrence and other potential factors, such as Gross Domestic Product (GDP) and geographic locations. Seventeen municipal areas were selected for this study—and only cities with more than 200 patients were considered. These cities covered a very large area with a combined population of 91.6 million and an area of 146,000 km^2. Geographical information for 54,351 cases from the chosen 17 areas was collected. Considering the IM incidence rate of the 17 analyzed municipal areas (Table 2), Xuzhou City had the lowest rate, 10.47%, while Changzhou City had the highest rate, 24.22%. From east to west and from south to north, IM incidence rate decreased stepwise. In the relationship between IM incidence and GDP per capita, IM incidence and *H. pylori* infection

Table 2 IM incidence, *H. pylori* infection, and GDP per capita in 17 prefecture-level cities

Area (province)	City	IM (n)	Biopsy (n)	IM rate (%)	*H. pylori* infection rate (%)	GDP (10,000RMB)
JiangSu	Xuzhou	29	277	10.47	28.88	19,480
	Suqian	143	1180	12.12	20.68	11,644
	Huaian	377	2735	13.78	23.69	16,031
	Lian-yungang	57	391	14.58	19.44	14,600
	Yancheng	253	1397	18.11	24.62	18,079
	Taizhou	166	781	21.25	22.02	25,085
	Yangzhou	192	943	20.36	25.03	29,036
	Wuxi	66	282	23.40	29.79	64,529
	Nantong	79	342	23.10	26.61	28,069
	Nanjing	8533	38,250	22.31	34.18	43,888
	Zhenjiang	355	1478	24.02	29.03	42,538
	Changzhou	187	772	24.22	29.02	43,833
AnHui	Chuzhou	553	3133	17.65	24.29	10,535
	Hefei	43	226	19.03	30.53	27,342
	Chaohu	282	1462	19.29	27.77	9760
	Xuancheng	44	211	20.85	30.33	12,960
	Ma-anshan	84	391	21.48	27.11	38,231

had similar geographical distributions (Data sources: Global CNKI, http://tongji.cnki.net/kns55/index.aspx). Linear regression analysis was performed to evaluate the correlation among the above factors, and the results indicated a high correlation between IM incidence and GDP per capita ($r = 0.67$; $P = 0.0030$) and between IM incidence and *H. pylori* infection ($r = 0.51$; $P = 0.0385$). We also found that *H. pylori* prevalence was higher in high-income populations.

Conclusion

Some risk factors discussed in this chapter have been investigated in previous studies, especially in GC research. Our study involved an unprecedented population size, and the large sample size gave us sufficient statistical power to not only validate conclusions of previous studies but also to reveal some new discoveries (Table 1). In this study, those reported factors with positive correlation to IM incidence were validated, such as age, GU, *H. pylori* infection, AG, dysplasia, and GC. DU is the only factor with a significant negative relation to IM. As a potential risk factor of IM, bile reflex was rarely mentioned in previous literature. In our study, bile reflex not only showed a positive relation with IM incidence, but also had an increase of IM level; furthermore, the percentage of patients with bile reflex increased as well. Although many studies were trying to access the risk factors of IM, some gaps still remain in the understanding of IM. A large-scale, long-term study is required to reveal relationships between *H. pylori* infection/eradication, IM, GC, and other risk factors. Through research and analysis, we systematically studied the IM status of southeastern China. Incidence of IM showed a regional characteristic distribution and was consistent with other reports of *H. pylori* infection and income distribution. In the future, we plan to integrate more data, such as regional characteristics distribution, climate, lifestyles, eating habits, and economic status—all of which will be helpful to clarify the pathogenesis of IM and the relationship between IM and GC.

Summary

Nowadays, "Precision Medicine" has become a new buzz word within the medical and research community, after President Obama's recent announcement of the Precision Medicine Initiative. In the next 50 years, this personalized and precision medicine might save hundreds of billions of dollars and prevent/treat disease much more effectively to improve the health in the US [18]. Without the support of big data in health informatics, personalized precision medicine could remain a theoretical hypothesis forever. Big data will be the key to accelerating the progress of translating the theoretical models to clinical applications in precision medicine. Although with the tremendous potential of precision medicine and its effect on health informatics, we are also facing big data challenges in other areas. The challenges are not limited to volume. They also involve the quality of data, which is difficult to identify, the speed of data access and connectivity, and the requirement of infrastructure to adapt to the explosion of incoming data. Finally, the last challenge, and perhaps the most important one, is finding the right talents and methods to discover the meaningful insights and interpret them. To conquer above obstacles, it requires not only an evolution of infrastructure, but also more initiative ideas for leaning big health informatics data with computational methods. Beyond that, collaborations among experts in different areas are also very important to address the challenges.

References

1. A. Acharya, J.J. VanWormer, S.C. Waring, A.W. Miller, J.T. Fuehrer, G.R. Nycz, Regional epidemiologic assessment of prevalent periodontitis using an electronic health record system. Am. J. Epidemiol. **177**(7), 700–707 (2013)
2. U.D. Akavia, O. Litvin, J. Kim, F. Sanchez-Garcia, D. Kotliar, H.C. Causton, P. Pochanard, E. Mozes, L.A. Garraway, D. Pe'er, An integrated approach to uncover drivers of cancer. Cell **143**(6), 1005–1017 (2010)

3. M.J. Boot, C.H. Westerberg, J. Sanz-Ezquerro, J. Cotterell, R. Schweitzer, M. Torres, J. Sharpe, In vitro whole-organ imaging: 4D quantification of growing mouse limb buds. Nat. Methods **5**(7), 609–612 (2008)
4. Z. Bu, J. Ji, A current view of gastric cancer in China. Transl. Gastrointest. Cancer 1–4 (2013)
5. Cancer Genome Atlas Network, Comprehensive molecular portraits of human breast tumours. Nature **490**(7418), 61–70 (2012)
6. Cancer Genome Atlas Research Network, Comprehensive genomic characterization defines human glioblastoma genes and core pathways. Nature **455**(7216), 1061–1068 (2008)
7. Cancer Genome Atlas Research Network, Comprehensive molecular characterization of gastric adenocarcinoma. Nature **513**(7517), 202–209 (2014)
8. R.M. Cantor, K. Lange, J.S. Sinsheimer, Prioritizing GWAS results: a review of statistical methods and recommendations for their application. Am. J. Hum. Genet. **86**(1), 6–22 (2010)
9. E. Cerami, J. Gao, U. Dogrusoz, B.E. Gross, S.O. Sumer, B.A. Aksoy, A. Jacobsen, C.J. Byrne, M.L. Heuer, E. Larsson, Y. Antipin, B. Reva, A.P. Goldberg, C. Sander, N. Schultz, The cBio cancer genomics portal: an open platform for exploring multidimensional cancer genomics data. Cancer Discov. **2**(5), 401–404 (2012)
10. A.T. Chan, S. Ogino, C.S. Fuchs, Aspirin and the risk of colorectal cancer in relation to the expression of COX-2. N. Engl. J. Med. **356**(21), 2131–2142 (2007)
11. ENCODE Project Consortium, An integrated encyclopedia of DNA elements in the human genome. Nature **489**(7414), 57–74 (2012)
12. A.C. de Vries, N.C. van Grieken, C.W. Looman, M.K. Casparie, E. de Vries, G.A. Meijer, E.J. Kuipers, Gastric cancer risk in patients with premalignant gastric lesions: a nationwide cohort study in the Netherlands. Gastroenterology **134**(4), 945–952 (2008)
13. Y. Demchenko, P. Grosso, C. de Laat, P. Membrey, Addressing big data issues in scientific data infrastructure, in *2013 International Conference on Collaboration Technologies and Systems (CTS)* (IEEE, 2013)
14. C.M. den Hoed, B.C. van Eijck, L.G. Capelle, H. van Dekken, K. Biermann, P.D. Siersema, E.J. Kuipers, The prevalence of premalignant gastric lesions in asymptomatic patients: predicting the future incidence of gastric cancer. Eur. J. Cancer **47**(8), 1211–1218 (2011)
15. J.C. Denny, Chapter 13: Mining electronic health records in the genomics era. PLoS Comput. Biol. **8**(12), e1002823 (2012)
16. M.F. Dixon, R.M. Genta, J.H. Yardley, P. Correa, Classification and grading of gastritis. The updated Sydney System. International Workshop on the Histopathology of Gastritis, Houston 1994. Am. J. Surg. Pathol. **20**(10), 1161–1181 (1996)
17. R.H. Duerr, K.D. Taylor, S.R. Brant, J.D. Rioux, M.S. Silverberg, M.J. Daly, A.H. Steinhart, C. Abraham, M. Regueiro, A. Griffiths, T. Dassopoulos, A. Bitton, H. Yang, S. Targan, L.W. Datta, E.O. Kistner, L.P. Schumm, A.T. Lee, P.K. Gregersen, M.M. Barmada, J.I. Rotter, D.L. Nicolae, J.H. Cho, A genome-wide association study identifies IL23R as an inflammatory bowel disease gene. Science **314**(5804), 1461–1463 (2006)
18. V. J. Dzau, G. S. Ginsburg, K. Van Nuys, D. Agus, D. Goldman, Aligning incentives to fulfil the promise of personalised medicine. The Lancet **385**(9982), 2118–2119 (2015)
19. C. Felley, H. Bouzourene, M.B. VanMelle, A. Hadengue, P. Michetti, G. Dorta, L. Spahr, E. Giostra, J.L. Frossard, Age, smoking and overweight contribute to the development of intestinal metaplasia of the cardia. World J. Gastroenterol. **18**(17), 2076–2083 (2012)
20. D. Garezs, M. Davis, Electronic Patient Records. EMRs and EHRs. Concepts as different as apples and oranges at least separate names. Health Informatics online (2005)
21. C.A. Gonzalez, M.L. Pardo, J.M. Liso, P. Alonso, C. Bonet, R.M. Garcia, N. Sala, G. Capella, J.M. Sanz-Anquela, Gastric cancer occurrence in preneoplastic lesions: a long-term follow-up in a high-risk area in Spain. Int. J. Cancer **127**(11), 2654–2660 (2010)
22. C.S. Greene, J. Tan, M. Ung, J.H. Moore, C. Cheng, Big data bioinformatics. J. Cell. Physiol. **229**(12), 1896–1900 (2014)
23. L. Han, Y. Yuan, S. Zheng, Y. Yang, J. Li, M.E. Edgerton, L. Diao, Y. Xu, R.G. Verhaak, H. Liang, The Pan-Cancer analysis of pseudogene expression reveals biologically and clinically relevant tumour subtypes. Nat. Commun. **5**, 3963 (2014)

24. G.E. Hoffman, B.A. Logsdon, J.G. Mezey, PUMA: a unified framework for penalized multiple regression analysis of GWAS data. PLoS Comput. Biol. **9**(6), e1003101 (2013)
25. E.S. Huang, L.L. Strate, W.W. Ho, S.S. Lee, A.T. Chan, Long-term use of aspirin and the risk of gastrointestinal bleeding. Am. J. Med. **124**(5), 426–433 (2011)
26. D.J. Hunter, P. Kraft, K.B. Jacobs, D.G. Cox, M. Yeager, S.E. Hankinson, S. Wacholder, Z. Wang, R. Welch, A. Hutchinson, J. Wang, K. Yu, N. Chatterjee, N. Orr, W.C. Willett, G.A. Colditz, R.G. Ziegler, C.D. Berg, S.S. Buys, C.A. McCarty, H.S. Feigelson, E.E. Calle, M. J. Thun, R.B. Hayes, M. Tucker, D.S. Gerhard, J.F. Fraumeni Jr., R.N. Hoover, G. Thomas, S. J. Chanock, A genome-wide association study identifies alleles in FGFR2 associated with risk of sporadic postmenopausal breast cancer. Nat. Genet. **39**(7), 870–874 (2007)
27. P.B. Jensen, L.J. Jensen, S. Brunak, Mining electronic health records: towards better research applications and clinical care. Nat. Rev. Genet. **13**(6), 395–405 (2012)
28. E.J. Kuipers, Helicobacter pylori and the risk and management of associated diseases: gastritis, ulcer disease, atrophic gastritis and gastric cancer. Aliment. Pharmacol. Ther. **11** (Suppl 1), 71–88 (1997)
29. A. Labrinidis, H. Jagadish, Challenges and opportunities with big data. Proc. VLDB Endowment **5**(12), 2032–2033 (2012)
30. P. Lemey, A. Rambaut, T. Bedford, N. Faria, F. Bielejec, G. Baele, C.A. Russell, D.J. Smith, O.G. Pybus, D. Brockmann, M.A. Suchard, Unifying viral genetics and human transportation data to predict the global transmission dynamics of human influenza H3N2. PLoS Pathog. **10** (2), e1003932 (2014)
31. S. Li, F. Garrett-Bakelman, A.E. Perl, S.M. Luger, C. Zhang, B.L. To, I.D. Lewis, A.L. Brown, R.J. D'Andrea, M. Ross, R. Levine, M. Carroll, A. Melnick, C.E. Mason, Dynamic evolution of clonal epialleles revealed by methclone. Genome Biol. **15**(9), 472 (2014)
32. J. Listgarten, C. Lippert, C.M. Kadie, R.I. Davidson, E. Eskin, D. Heckerman, Improved linear mixed models for genome-wide association studies. Nat. Methods **9**(6), 525–526 (2012)
33. B.A. Logsdon, G.E. Hoffman, J.G. Mezey, A variational Bayes algorithm for fast and accurate multiple locus genome-wide association analysis. BMC Bioinform. **11**, 58 (2010)
34. J. Lurio, F.P. Morrison, M. Pichardo, R. Berg, M.D. Buck, W. Wu, K. Kitson, F. Mostashari, N. Calman, Using electronic health record alerts to provide public health situational awareness to clinicians. J. Am. Med. Inform. Assoc. **17**(2), 217–219 (2010)
35. S.G. Megason, S.E. Fraser, Imaging in systems biology. Cell **130**(5), 784–795 (2007)
36. J.J. Michaelson, Y. Shi, M. Gujral, H. Zheng, D. Malhotra, X. Jin, M. Jian, G. Liu, D. Greer, A. Bhandari, W. Wu, R. Corominas, A. Peoples, A. Koren, A. Gore, S. Kang, G.N. Lin, J. Estabillo, T. Gadomski, B. Singh, K. Zhang, N. Akshoomoff, C. Corsello, S. McCarroll, L. M. Iakoucheva, Y. Li, J. Wang, J. Sebat, Whole-genome sequencing in autism identifies hot spots for de novo germline mutation. Cell **151**(7), 1431–1442 (2012)
37. L. Olsen, J.M. McGinnis, *Redesigning the Clinical Effectiveness Research Paradigm: Innovation and Practice-Based Approaches: Workshop Summary* (National Academies Press, 2010)
38. J. Pathak, R.C. Kiefer, C.G. Chute, Using linked data for mining drug-drug interactions in electronic health records. Stud. Health Technol. Inform. **192**, 682–686 (2013)
39. B. Peleteiro, N. Lunet, C. Figueiredo, F. Carneiro, L. David, H. Barros, Smoking, *Helicobacter pylori* virulence, and type of intestinal metaplasia in Portuguese males. Cancer Epidemiol. Biomark. Prev. **16**(2), 322–326 (2007)
40. A. Pollock, S. Bian, C. Zhang, Z. Chen, T. Sun, Growth of the developing cerebral cortex is controlled by microRNA-7 through the p53 pathway. Cell Rep. **7**(4), 1184–1196 (2014)
41. K. Sakitani, Y. Hirata, H. Watabe, A. Yamada, T. Sugimoto, Y. Yamaji, H. Yoshida, S. Maeda, M. Omata, K. Koike, Gastric cancer risk according to the distribution of intestinal metaplasia and neutrophil infiltration. J. Gastroenterol. Hepatol. **26**(10), 1570–1575 (2011)
42. N. Savage, Bioinformatics: big data versus the big C. Nature **509**(7502), S66–S67 (2014)

43. S.J. Shah, D.H. Katz, S. Selvaraj, M.A. Burke, C.W. Yancy, M. Gheorghiade, R.O. Bonow, C.C. Huang, R.C. Deo, Phenomapping for novel classification of heart failure with preserved ejection fraction. Circulation **131**(3), 269–279 (2015)
44. P. Sipponen, M. Kekki, M. Siurala, Age-related trends of gastritis and intestinal metaplasia in gastric carcinoma patients and in controls representing the population at large. Br. J. Cancer **49**(4), 521–530 (1984)
45. G.W. Sledge Jr., R.S. Miller, R. Hauser, CancerLinQ and the future of cancer care. Am. Soc. Clin. Oncol. Educ. Book 430–434
46. M. Stolte, A. Meining, The updated Sydney system: classification and grading of gastritis as the basis of diagnosis and treatment. Can. J. Gastroenterol. **15**(9), 591–598 (2001)
47. T. Tsukui, R. Kashiwagi, M. Sakane, F. Tabata, T. Akamatsu, K. Wada, S. Futagami, K. Miyake, N. Sueoka, T. Hirakawa, M. Kobayashi, T. Fujimori, C. Sakamoto, Aging increases, and duodenal ulcer reduces the risk for intestinal metaplasia of the gastric corpus in Japanese patients with dyspepsia. J. Gastroenterol. Hepatol. **16**(1), 15–21 (2001)
48. X. Wang, Z. Duren, C. Zhang, L. Chen, Y. Wang, Clinical data analysis reveals three subtypes of gastric cancer, in *2012 IEEE 6th International Conference on Systems Biology (ISB)* (IEEE, 2012)
49. H. Xie, M. Wang, A. de Andrade, F. Bonaldo Mde, V. Galat, K. Arndt, V. Rajaram, S. Goldman, T. Tomita, M.B. Soares, Genome-wide quantitative assessment of variation in DNA methylation patterns. Nucleic Acids Res. **39**(10), 4099–4108 (2011)
50. C. Zhang, S. Xu, D. Xu, Risk assessment of gastric cancer caused by *Helicobacter pylori* using CagA sequence markers. PLoS ONE **7**(5), e36844 (2012)

Computational Infrastructure for Telehealth

Fedor Lehocki, Igor Kossaczky, Martin Homola and Marek Mydliar

Abstract Recent developments in society show significant trends in aging population and prevalence of chronic conditions. It is estimated that disruptive demographics related to population of middle-aged and older adults will result in 33% of overall EU population by 2025. Advances in technology created innovative means to support effective management of these challenges through telehealth model for healthcare delivery. In this chapter we introduce decision support in hypertension management with ontology describing the structure of the relevant domain data and analysis of such data using a rule-based system. Telehealth solution provides a 'complete-loop' concept for hypertension management with sensor device, mobile, and web-based applications providing means for health status management for both healthcare consumer and healthcare provider.

Introduction to Telehealth

Recent developments in society show two significant trends reflecting in aging population and prevalence of chronic conditions. It is estimated that disruptive demographics related to population of middle-aged and older adults will result in 33% of overall EU population by 2025. This is closely associated with increase of

F. Lehocki (✉) · I. Kossaczky · M. Mydliar
National Centre of Telemedicine Services, Faculty of Electrical Engineering and Information Technology, Slovak University of Technology in Bratislava, Bratislava, Slovakia
e-mail: fedor.lehocki@stuba.sk

I. Kossaczky
e-mail: igor.kossaczky@stuba.sk

M. Mydliar
e-mail: marek.mydliar@mhealth.sk

M. Homola
Faculty of Mathematics, Physics and Informatics (FMFI), Comenius University in Bratislava (UK), Bratislava, Slovakia
e-mail: homola@fmph.uniba.sk

D. Xu et al. (eds.), *Health Informatics Data Analysis*, Health Information Science,
DOI 10.1007/978-3-319-44981-4_12

patients with chronic conditions—CVDs, diabetes, COPD, cancer, epilepsy, arthritis, asthma, obesity, and overweight [1]. Approximately 90% of seniors have at least one chronic disease, and 77% have two or more chronic conditions [2]. Based on the data from WHO there are 347 million people with diabetes worldwide and this number will double by 2030. CVDs represent number one cause of death globally, more people die annually from these health conditions than from any other cause [3]. Despite the fact that chronic diseases are among the most common and costly health problems, they are also among the most preventable and most can be effectively controlled.

Advances in technology that we have been witnessing for the past decade created innovative means to support effective management of the challenges related to chronic disease management, health and wellness and aging independently through telehealth model for healthcare delivery [4]. This includes broadband Internet access (with its adoption by senior citizens), cheap mobile technologies, smartphones, miniaturization of sensors, wearable systems (e.g., based on wireless technology and e-textiles), development of MEMS (including low power consumption—accelerometers, gyroscopes, magnetometers), and off-shelf personal health monitoring devices. By definition of American Telemedicine Association telehealth is the use of medical information exchanged from one site to another via electronic communications to improve a patient's clinical health status. Telehealth and telemedicine can be considered as interchangeable terms when addressing a wider definition of remote healthcare with respect to remote patient monitoring, referral specialist services in primary care, consumer medical and health information, and medical education.

Telehealth applications such as mobile health monitoring, remote ICU, remote ECG monitoring, teledermatology, and teleradiology generate heterogeneous biomedical data. Detailed analysis for some of these data (EEG, ECG, image processing) is covered in several chapters of this book. We will consider computational infrastructure of telehealth from the perspective of so-called 'completing the loop' related to remote data collection, data transmission, expert review, and feedback [5]. In the next sections we will illustrate this loop with underlying technologies, challenges and references to the state-of-the-art computer programs that can contribute to creation of modern health care through telehealth services.

Architecture of Telehealth Systems

Overview

Telehealth solution that empowers better management of health and wellness may consist of the following basic components (see Fig. 1): personal health devices (PHD), gateway device (GD), and remote patient monitoring server (RPM).

Personal health device: a sensor that measures individual's vital parameters or daily activities [7]. These are low-powered devices with few processing resources

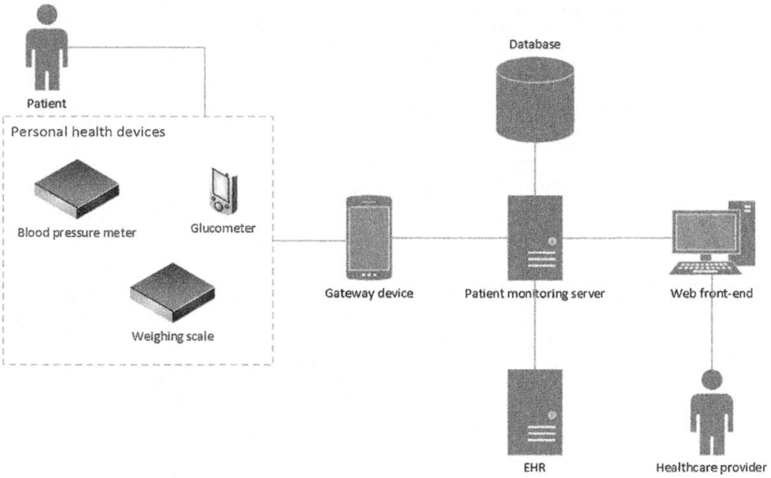

Fig. 1 Personal telehealth system architecture

that send data to the gateway device. Examples of communication interface between PHD and GD include USB, Bluetooth, Bluetooth Low Energy, ZigBee, ANT+, and WiFi. More detailed information about sensing and wearable devices for telehealth solutions can be found in [6].

Gateway device: the residential gateway and central point of control that include cell phones (smartphones), personal computers, set-top boxes or other special devices (e.g., designed for elderly people with larger display and function buttons). Device is responsible for collecting measurements from the different PHDs and forwarding them to the remote patient monitoring server. Examples of communication interface include WiFi, or cellular networks. Some of device functionalities are local data storage, basic data analysis and graphical user interface for viewing actual and historical patient's measurements and management of other contextual health-related data (text messages containing additional info about patient's health status, level of patient's discomfort, etc.).

Remote patient monitoring server (RPM): this is a backend system of telehealth service provider that enables healthcare professionals to view, analyze, and manage patient's data. The user interface is implemented as web front-end irrelevant on the device that healthcare professional uses (e.g., laptop, desktop computer, mobile devices). Service platform is responsible for storing data in a persistent database. Data can be also forwarded to other components and systems (EHR server at hospitals or other primary care locations).

Components of personal telehealth systems are supplied by a variety of vendors that come with proprietary communication protocols and data encodings. Although this seems reasonable regarding the acceptable development complexity and initial

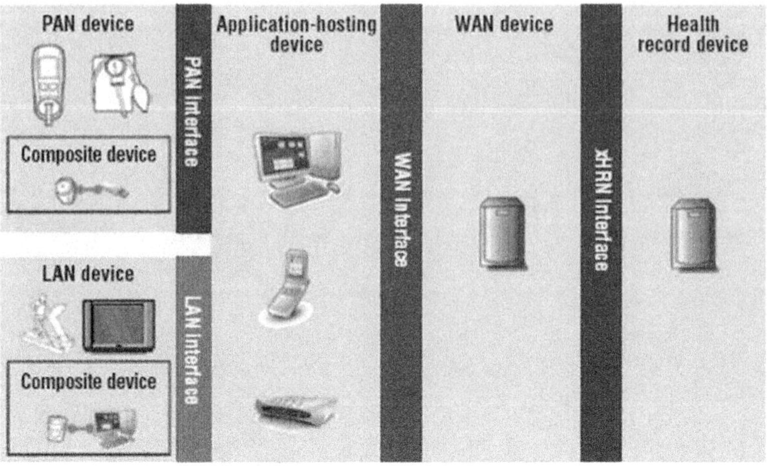

Fig. 2 Continua end-to-end reference architecture

implementation costs it restricts creation of plug and play telehealth solutions and is one of the reasons that prevent their wider acceptance. The objective is to ensure interoperability so that a single system is able to collect, process, and analyze data by various personal health devices, gateway devices, and RPMs.

Continua Health Alliance (CHA) is a nonprofit, open industry organization of more than 200 leading healthcare and technology companies joining together in collaboration dedicated to establishing a system of interoperable personal connected health solutions [7, 8]. The ISO/IEEE Std 11073-20601 and ISO/IEEE Std 11073-20601A-20101 Personal Health Device standards [9] are recommended by CHA to address the PHD and GD interoperability (Fig. 2). The transport layer is profiled for Bluetooth, Bluetooth Low Energy, ZigBee, and USB technology. IEEE standards define messages that travel between PHD (called agent in IEEE terminology or PAN, LAN device in Continua terminology), and gateway device (called manager in IEEE terminology or Application hosting device in Continua terminology). The agent is modeled as a set of objects. Object attributes represent measurements and status data that are sent to a manager. Presenting its object structure (configuration) PHD supports plug and play interoperability.

The object model methodology allows a unique generic mapping to IHE-PCD01 (HL7)[1] messages. These can be further transported by the GD to the RPM server (called WAN Device in Continua terminology) using a number of transport methods over Internet and private TCP/IP networks including secured web service interfaces. Patient information can be forwarded from RPM to Electronic Health Record systems (EHR, called Health Reporting Network Device in Continua

[1]Integrating the Healthcare Enterprise (IHE) is an initiative by healthcare professionals and industry to improve the way computer systems in healthcare share information. It promotes cooridnated use of established standards such as Health Level 7 (HL7).

terminology). The recommended standards for establishing this communication are IHE XDR profile and HL7 Personal Health Monitoring Report document (PHM) [10]. By passing the Continua certification process vendors assure that their components can be interconnected and can exchange data.

Security and privacy related to collection and processing of personal and health data is very strictly regulated in number of countries through legislation such as EU Directive 95/46 and HIPAA in the US. In its guidelines CHA considered advanced security and privacy requirements such as identity management, non-repudiation of origin, and consent management. More details on the topic of data structure, management, privacy, and security of telehealth infrastructure can be found in Continua Design Guidelines [11].

Information about other available certified devices for interoperable state-of-the-art telehealth solutions can be found on the Continua web site.[2]

Data Analytics

Expert review and feedback in telehealth systems can be supported by Clinical Decision Support Systems (CDSS) that may compete with the increasing load of clinical data by providing integrated approach to their analyses. These systems provide information management help focusing clinician's attention, foster adherence to guidelines, prevent mistakes, provide patient-specific recommendations, and spread up specialist knowledge to primary care clinicians. For example, follow-up and early detection of heart failure patient's decompensation can be supported by integration of signal and image processing methods by means of a CDSS [12]. Several architectures have been indicated for CDSS linked with other systems including stand alone and decision support systems integrated with clinical systems like EHR [13]. Comprehensive taxonomy of clinical decision support tools for major nine commercial and four internally developed EHR systems were evaluated in [14]. Decision support capabilities included medication dosing support, order facilitators, point-of-care alerts/reminders, relevant information display, expert systems, and workflow support. Table 1 summarizes surveyed vendors and institutions.

With the aim to avoid unnecessary further generalization, we introduce example of decision support for hypertension management based on symbolic paradigm formalizing expert's knowledge into ontology and rule-based knowledge base. In knowledge engineering, ontologies [15] are used to formalize shared understanding of a domain through definition of concepts and relationships among them. This enables both software applications and humans to share the knowledge related to a domain of interest. Ontologies can be seen as semantic vocabularies used to model a domain providing a set of general categories for classification of data (classes),

[2]http://www.continuaalliance.org/.

Table 1 Vendors and institutions

Vendor	**Product name**
Allscripts	Allscripts EHR
Cerner	PowerChart/PowerWorks
Eclipsys	Sunrise Clinical Manager
e-MDs	Solution Series
Epic	EpicCare Inpatient
NextGen	Inpatient Clinicals
GE	Centricity EMR
GMT	PrimeSuite
Springcharts	Springcharts EHR
Institution	**Product name**
Partners Healthcare	LMR
Veteran's affairs Health System	VistA
Regenstrief Institute	RMRS
Intermountain Healthcare	HELP-2

and their relations (properties). Objects of certain domain (instances) can be classified in the ontology (i.e., assigned into classes) and interrelated using the properties. An important feature of ontologies is that the meaning of each class and of each property is precisely defined. Ontologies can be complemented by other representation formalisms such as rules that are the most suitable for expressing single medical decisions (e.g., alerts occurring in case of significant changes in values of measured physiological signals) [16].

Knowledge management platforms which can be used to craft decision support on top of domain data include ontology editors such as Protégé[3]; ontologically annotated data can be stored in semantic data stores such as OWLIM[4]; basic interface between ontologies and rules can be provided, e.g., by the Apache JENA platform[5]; more complex rule engines include XSB[6] and DLV.[7]

Use Case: Telehealth Solution for Hypertension Management

In this section we introduce an example use case for decision support in hypertension management. In Section "Data Structure and Analytics" we start from a proposed ontology describing the structure of the relevant domain data. We then

[3]http://protege.stanford.edu.

[4]http://www.ontotext.com/owlim.

[5]http://jena.apache.org.

[6]http://xsb.sourceforge.net.

[7]http://www.dlvsystem.com.

briefly show how analysis of such data (in the structure prescribed by the ontology) can be done using a rule-based system. Section "Implementation of Telehealth System for Hypertension Management" covers the implemented telehealth solution based on this data architecture, which provides a 'complete-loop' telehealth solution concept for hypertension management with sensors, mobile and web based applications providing means for health status management for both healthcare consumer and healthcare provider.

Data Structure and Analytics

For the needs of our use case we have developed a simple domain ontology, as depicted in Fig. 3. Three classes are central in the ontology: Class *Patient* encapsulates data respective to a patient. Class *Measurement_Limit* and its subclasses capture limit values for various measurements. Class *Reading* captures readings from a measuring device and data about patient subjective feelings (type and level of discomfort, and stress). Readings from measuring device include time stamp of measurement, pulse rate, systolic and diastolic blood pressure, and arrhythmia indication.

Class *Patient* (see Fig. 4a) groups information about patients. Each patient has a name and can be associated with one or more readings, one or more diagnosis, and possibly some instances of (different subclasses of) *Measurement_Limit*. Note that the measurement limits are associated via functional properties (i.e., only one limit of each type can be associated). The instances of *Measurement_Limit* define threshold values for various measurements related to the acquisition of patients' health status. These instances (see Fig. 4b) are strictly split into two subclasses *Personal_Measurement_Limit* (specific limits defined for a particular patient) and *Generic_Measurement_Limit* (default limits for patients who don't have the specific limit set). Instances of *Generic_Measurement_Limit* class store threshold values for measurements based on diagnoses. In case there is no specific diagnosis associated with patient the instance stores a default value. It is important to note that personalization of threshold values through instances of *Personal_Measurement_Limit* is given the highest priority.

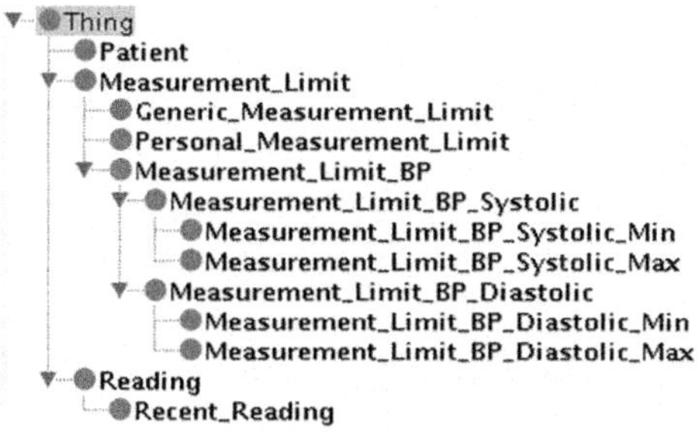

Fig. 3 Domain ontology (class hierarchy)

(a) Class Patient:

Patient
- has_name xsd:String (required)
- has_reading Reading
- has_diagnosis xsd:String
- has_limit_systolic_min Measurement_Limit_BP_Systolic_Min (functional)
- has_limit_systolic_max Measurement_Limit_BP_Systolic_Max (functional)
- has_limit_diastolic_min Measurement_Limit_BP_Diastolic_Min (functional)
- has_limit_diastolic_max Measurement_Limit_BP_Diastolic_Max (functional)

(b) Class Measurement_Limit:

Measurement_Limit
  = disjoint union of Generic_Measurement_Limit
    and Personal_Measurement_Limit
- value xsd:Real (required)

Generic_Measurement_Limit
  (possibly assoc. with multiple Patient instances)
- for_diagnosis xsd:String

Personal_Measurement_Limit
  (associated with single Patient instance)

(c) Class Reading:

Reading
- timestamp xsd:DateTime (required)
- pulse rate xsd:Real
- systolic_pressure xsd:Real
- diastolic_pressure xsd:Real
- arrhytmia xsd:Boolean
- discomfort – xsd:String
- discomfortLevel xsd:Integer range[0..10]
- stress – xsd:Boolean

Fig. 4 Domain ontology (details)

Independently, the *Measurement_Limit* class splits into specific subclasses (e.g., *Measurement_Limit_BP_Systolic*, *Measurement_Limit_BP_Diastolic*) indicative of the specific limit type. Further subclasses indicate minimal and maximal limits. See Fig. 3 for the full overview. The actual value of the limit is given by the *value* property.

Instances of *Reading* (see Fig. 4c) encapsulate data about patients' subjective feelings and data received from a measuring (sensor) device. All values measured and reported at the same time are stored by the system as one *Reading* instance. The readings which occurred recently are further classified under the *Recent_Read*ing subclass.

The ontology is used as a schema for the application data stored in a *fact base* (e.g., the information that a patient John Smith has diagnosis "chronic kidney disease" is stored using the facts: *Patient* (*p555*), *has_name* (*p555*, "John Smith"), and *has_diagnosis* (*p555*, "chronic kidney disease"). Data in the fact base is then evaluated with a rule engine. Production rules (stored in the *rule base*) have the simple form: IF <*condition*> THEN <*action*>.

The rule engine is activated upon addition of new data in the fact base and rules whose <*condition*> matches the facts are fired. The execution is over when no matching rule can be found anymore. Let us see, for example, the following rules which are responsible for associating a patient with the maximum limit of systolic pressure:

(1) IF

Patient(?P) AND not *has_limit_systolic_max*(?P,_)

AND *Measurement_Limit_BP_ Systolic_Max*(?M)
AND *Generic_Measurement_Limit*(?M)
AND not *for_diagnosis*(?M,_)

THEN

store(*has_limit_systolic_max*(?P,?M))

(2) IF

Patient(?P) AND *has_limit_systolic_max*(?P,?M)

AND *Generic_Measurement_Lim*it(?M)
AND *has_diagnosis*(?P, "chronic kidney disease")
AND *Measurement_Limit_BP_Systolic_Max*(?O)
AND *Generic_Measurement_Limit*(?O)
AND *for_diagnosis*(?O, "chronic kidney disease")

THEN

store(*has_limit_systolic_max*(?P,?O))

General objective of Rules (1) and (2) is to always have the most appropriate assignment of maximum limit for systolic pressure associated with each patient. If nothing further is known about the patient a default value that is not associated with any specific diagnosis (stored in the fact base itself) should be linked to the patient.

Rule (1) is responsible for association of default limit of maximum systolic blood pressure to a patient (in case a patient does not have any other limit already associated).

It fires whenever the variables ?P and ?M can be bound with a patient with no limit yet assigned and with the generic default limit, respectively. A new fact about these bound objects is stored in the fact base in the rule's action statement. Note that the sign "_" is a placeholder that matches with any value.

Rule (2) fires in case of patients diagnosed with "chronic kidney disease". The rule re-associates any such patient with the more specific default systolic limit related to this diagnosis.

Note that neither rule (1) or (2) fires for patients for which the most specific personal limit was already stored in the fact base (by a new *Personal_ Measurement_Limit* (L) set to a respective value and associating it with the patient). This is because in such a case *Generic_Measurement_Limit* (L) is not true. The analysis of current readings for each patient is then done by the following rule, which fires if a recent systolic pressure reading respective to some patient is found and compares its value with the limit associated with the patient:

(3) IF

Patient(?P) AND *has_name*(?P,?N)

AND *has_limit_systolic_max*(?P,?L)
AND *value*(?L,?X)
AND *Recent_Reading*(?R)
AND *has_reading*(?P,?R)
AND *systolic_pressure*(?R,?Y)
AND ?Y > ?X

THEN

exec(*Alert*("High sistolic pressure ?N"))

As a result of Rule (3) an alert is executed if the actual reading surpassed the limit value. Note that for simplicity we omit rules which compare the timestamps with current time and distinguish which readings are recent.

The explanation of the rule language and firing mechanism is illustratory and it is largely simplified in this chapter. For more details on rule-based systems refer to the literature (e.g., [17, 18, 19, 20]). For further examples of rule-based telemedicine systems see [21].

Implementation of Telehealth System for Hypertension Management

Based on consultations with clinicians we present simple telehealth solution for hypertension patients. It relies on the simplified version of personal telehealth system architecture shown on Fig. 1 and consists of a blood pressure sensor, mobile application implementing the functionality of a gateway device, and a web application illustrating the basic features of a remote patient monitoring server.

Patient health device is Boso-Medicus Prestige, the Bluetooth blood pressure meter that was chosen for the reliability and easiness of interface implementation with mobile application. Mobile application is running on Android OS (min. version 2.2.). It enables patient to enter his subjective situation and feelings (stress, discomfort—headache, dizziness,...) and to take blood pressure measurements that are wirelessly transmitted to the gateway device (systolic and diastolic pressure, pulse rate and arrhythmia indication). More details about blood pressure meter implementation can be obtained directly from the manufacturer (Boso) in its document related to communication protocol [22].

Readings are displayed and stored in local phone memory. By readings we understand measurements and other health-related information provided by patient. The GUI allows patient also to input measurements manually (for different type of sensor without Bluetooth connection). The historical data stored in phone memory can be viewed in a tabular or graphical form (see Fig. 5).

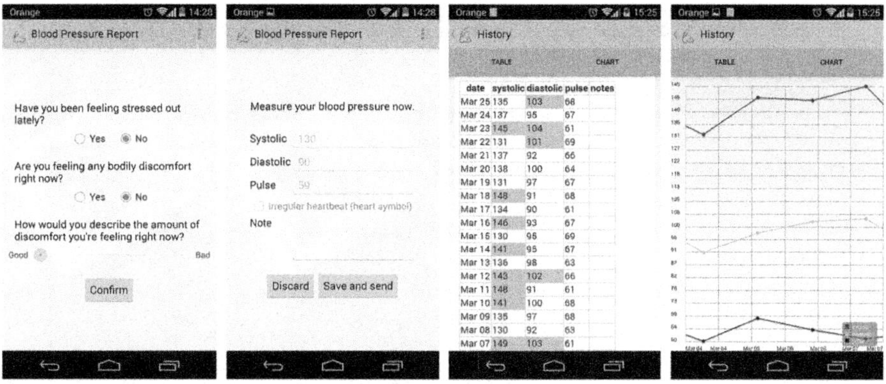

Fig. 5 Graphical User Interface of mobile application

Although the mobile application can be used in offline mode, i.e., without the web application, its main task is to forward readings to RPM where data can be reviewed by healthcare provider. The mobile application performs some basic data analytics. For each measured parameter, threshold levels (min., max.) can be defined so that application can generate alerts. The patient can set thresholds manually or download recommended personalized values by clinician from RPM.

Remote Patient Monitoring Server comprises three main software components:

(1) *RESTfull web-service* for client mobile application implements and publishes operations for:

- the patient registration with preferred healthcare provider,
- receiving, storing, and analyzing readings,
- downloading patient's personal thresholds for measured parameters.

(2) *Web application* with interface enable healthcare provider to perform the following activities:

- select a patient, view and edit his personal data and readings, view the history of readings in tabular and graphical form, Fig. 6a. View the alarm list displaying alerts from all his patients sorted by time. Alarms can be marked indicating that have been closed/seen by healthcare provider, Fig. 6b,
- print table of readings and alarm list.

(3) *Alarm generation module* uses a rule-based system to evaluate all incoming readings and generate alerts whenever rules' conditions are met. The basic processing calculates the daily average of systolic and diastolic pressure for each patient. If the average is above a defined limit threshold (3-day Max.) during three consecutive days, the high pressure 3-days-alarm is generated.

Fig. 6 **a** Chart displaying measurement history of a patient; **b** alarm list

Generic and personal, patient-specific limits can be defined and stored in a fact base as described in Section "Data Structure and Analytics". Healthcare provider can use web application dialog shown on Fig. 7a to configure both types of thresholds.

Besides the 3-days-alarms additional production rules for generation of isolated alarms can be defined using the dialog shown on Fig. 7b. An isolated alarm is associated with a single reading. Whenever a new reading is received it is instantly evaluated. If the rule condition is fulfilled the alarm is generated. A simple rule can have a form of comparison of a measured value with a limit (Min. or Max.):

$$systolic\_pressure > systolic\_pressure\_max$$

or evaluation of additional indicators describing patients feelings and situation. More complex condition can be composed using logical connectives (AND and OR):

Fig. 7 **a** Thresholds configuration; **b** Rules definition

$$\textit{systolic\_pressure} > \textit{systolic\_pressure\_max}\ \text{AND}$$
$$(\textit{discomfort}\ \text{contains}\ '\textit{Tightness in the chest}'\ \text{OR}\ \textit{discomfort}\ \text{contains}\ '\textit{headache}')$$

Lower and upper thresholds for isolated measurements are also defined in ontology and configured in the same way as limits for 3-days-alarms (Fig. 7a).

More detailed information about the telehealth system, mobile, and web demo applications and user's manual can be found on the web address http://www.monitor.mhealth.sk/.

Conclusion

In this chapter we provided an introduction to telehealth systems and overview of their computational architecture and challenges associated with creation of the 'complete-loop' solution. References to available computer programs and devices should provide the reader with practical tools to develop his own telehealth solutions. We have also included a practical use case describing simple application for monitoring patients with hypertension, with the aim to illustrate some major areas of knowledge management. This should help reader to further build the understanding of design and implementation of useful systems that can evolve with development of medical knowledge.

It is important to understand that besides technical challenges mentioned here there are also nontechnical barriers to adoption of telehealth systems. These are related to acceptance of new processes of healthcare provision and organizational change by clinicians, legal liability, reimbursement models, gathering of evidence regarding the impact of telehealth applications and patients and medical personnel education. Examples like 3 million lives[8] do exist that can serve as basic inspiration for further development of complex infrastructure for telehealth.

Acknowledgements This work was supported by project CARDINFO (APVV 0513/10) and ERDF projects SMART I (26240120005) and Competence Centre (26240220072).

References

1. Chronic diseases, World Health Organization, http://www.who.int/chp/en/, Retrieved Mar 25, 2014
2. National Center for Health Statistics, http://www.cdc.gov/nchs/hus.htm, Retrieved Mar 25, 2014
3. Fact Sheets from WHO Media Centre, (http://www.who.int/mediacentre/factsheets/en/), Retrieved Mar 25, 2014
4. F. Wartena, J. Muskens, L. Schmitt, M. Petković, Continua: The reference architecture of a personal telehealth ecosystem, 12th IEEE International Conference on e-Health Networking Applications and Services (Healthcom), (2010)
5. K.S. Nikita, (2013) in *Handbook of Biomedical Telemetry*, (Wiley-IEEE Press, 2013)
6. Y. Zheng, X. Ding, C. Poon, B. Lo, H. Zhang, X. Zhou, G.Z. Yang, N. Zhao, Y.T. Zhang, (2014) Unobtrusive sensing and wearable devices for health informatics. IEEE Trans. Biomed. Eng. **61**(5), 1538–1554 (2014)

[8] http://3millionlives.co.uk/.

7. M. Benner, L. Schope, in *12th IEEE International conference on Mobile Data Management*, Using Continua Health Alliance Standards, (2011)
8. R. Carroll et. al., in *IEEE Pervasive Computing*, Continua: an interoperable personal healthcare ecosystem, vol. 6, no. 4 (2007)
9. IEEE Health informatics–Personal health device communication Part 20601: Application profile-Optimized Exchange Protocol.–Amendment 1. *IEEE 11073-20601a-2010 (Amendment to IEEE Std 11073-20601-2008)*, vol., no., pp. 1,119, Jan. 24 2011
10. https://www.hl7.org/implement/standards/, Retrieved, 5 Apr 2014
11. Continua Design Guidelines, Version 2012 + Errata, 5 Nov 2012
12. F. Chiarugi et al., Biomedical signal and image processing for decision support in heart failure, in *MDA 2008, LNAI 5108* ed. by P. Perner, O. Salvetti (2008), pp. 38–51
13. A. Wright, D.F. Sittig, A four-phase model of the evolution of clinical decision support architectures. Int J med Inform. **77**, 641–649 (2008)
14. A. Wright, D.F. Sittig, J.S. Ash, Development and evaluation of a comprehensive clinical decision support taxonomy: comparison of front-end tools in commercial and internally developed electronic health record systems. J. Am. Med. Inform. Assoc. **18**, 232–242 (2011)
15. S. Staa , R. Studer (ed.), *Handbook on Ontologies* (Springer, 2004)
16. M. Peleg, S. Tu, in *International Medical Informatics Association (IMIA) Yearbook of Medical Informatics*, ed. by R. Haux and C. Kulikowski. Decision Support, Knowledge Representation and Management in Medicine: Assessing Information Technologies for Health, (Schattauer GmbH, Germany,2006)
17. D.A. Waterman, A guide to expert systems. (Addison-Wesley, 1986)
18. L. Sterling, E. Sapiro, *The Art of Prolog*. MIT Press, 2nd edn. (1994)
19. J.J. Carroll, I. Dickinson, C. Dollin, D. Reynolds, A. Seaborne, K. Wilkinson, In procs. of the 13th international World Wide Web Conference on Alternate track papers & posters, Jena: implementing the Semantic Web recommendations. (ACM, 2004)
20. M. Jang, J.C. Sohn, in *Rules and Rule Markup Languages for the Semantic Web*, Bossam: an extended rule engine for OWL inferencing. (Springer, 2004)
21. A. Minutolo, G. Sannino, M. Esposito, G. De Pietro, in *2010 IEEE EMBS Conference on Biomedical Engineering and Sciences (IECBES)*, A rule-based mHealth system for cardiac monitoring, (IEEE, 2010), pp. 144–149
22. Boso (Bosch + Son Germany, contact Mr. Harald Weigle), http://www.boso.de/boso/kontakt/d-jungingen-firmensitz.html, Retreived 25 Mar 2014

Healthcare Data Mining, Association Rule Mining, and Applications

Chih-Wen Cheng and May D. Wang

Abstract In this chapter, we first introduce data mining in general by summarizing popular data mining algorithms and their applications demonstrated in real healthcare settings. Afterward, we move our focus on a mining technique called association rule mining that can provide a more flexible data mining solution for personalized and evidence-based clinical decision support. Feasibility on how to use association rule mining is offered along with one example. The chapter concludes with a discussion of challenges that hamper the clinical use of conventional association rule mining and a few point-by-point solutions are provided.

Background of Data Mining Methods and Challenges

Advanced information technologies promise the massive influx of clinical- and person-centered data. These rich sources of data grant potential for an increased understanding of disease mechanisms and patient-centered decision-making so as to improve the quality of healthcare. However, the size, complexity, and biases of the data pose new challenges, which makes it difficult to transform the data to useful and actionable knowledge using conventional statistical analysis. Such a so-called "Big-Data" era raises an emerging and urgent need for scalable and computer-based data mining methods and tools that can discover useful patterns in a flexible, cost-effective, and productive way. In this section, we discuss three main data mining categories, including classification, clustering, and association rule mining [1].

M.D. Wang (✉)
Georgia Institute of Technology and Emory University, Atlanta, USA
e-mail: maywang@bme.gatech.edu

C.-W. Cheng
Wallace H. Coulter Department of Biomedical Engineering, Georgia Institute of Technology and Emory University, 313 Ferst Dr, UA Whitaker Bldg., Suite 4106, Atlanta, GA 30332, USA
e-mail: cwcheng83@gatech.edu

D. Xu et al. (eds.), *Health Informatics Data Analysis*, Health Information Science,
DOI 10.1007/978-3-319-44981-4_13

Classification

Classification is a common data mining method used to assign data into prespecified categories, called "classes" that usually represent an attribute in which users are most interested (e.g., suffering from disease vs. healthy). Classification is a supervised mining method since the class labels in the training dataset are provided. The goal of a classification algorithm is to create a model consisting of classification regulations; afterward, when new data (or records) is available, the model can accurately classify the data. For example, based on observed symptoms and clinical conditions, a well-trained classifier can help a provider to quantitatively determine (i.e., diagnose) a patient's cancer stage or predict prognosis, so as to provide proper treatment.

Common classification methods include Bayesian classifier [2], neural network [3], and decision tree [4]. Support vector machine (SVM) is another widely used classification method. SVM is used when a dataset is represented by two (i.e., binary) classes in a high-dimensional feature space. SVM searches for an optimal separating hyperplane as a decision boundary that maximizes the margin between two classes. To find the hyperplane with maximal margin, SVM uses support vector, and the margin is determined by using the two-support vector. Basic SVM utilizes linear kernel function as advanced SVM adopts nonlinear kernel function for better classification accuracy. The major advantage of SVM is its accuracy since it is proven to provide better performance than other classification methods [5]. Nevertheless, the performance highly depends on kernel selection, but selecting a right kernel function is a challenging issue. In addition, SVM is designed to resolve binary classification problems. A multi-class dataset should be first divided into multiple binary problems. Finally, compared with other classification, training step of SVM requires extremely high computational power, posing the user to tradeoff between performance and accuracy. Recent studies have adopted SVM in applications of chronic fatigue syndrome using genetic data [6], ovarian cancer using mass spectrometry data [7], and detecting abnormal activities in medical wireless sensor networks [8].

Clustering

Unlike classification that has a training dataset with predefined class labels, clustering, or unsupervised learning, refers to determining the hidden structure in an unlabeled dataset [9]. Clustering can be best used for the studies of large data of high dimensionality, but in which there is limited knowledge about the data. The goal of clustering models is to group data entries into a specific number of clusters so that the entries in each cluster share high similarity and entries from different clusters have low similarity.

Over the last few decades, studies have introduced a number of clustering algorithms. These algorithms are categorized in two main groups: agglomerative and partitional. Agglomerative clustering merges the most common groups on the basis of their pairwise similarities and forms a hierarchical structure. Based on different similarity measures, hierarchical clustering can be further divided into three types, including single-link, complete-link, and average-link, and among which average-link has been proven the best regarding the accuracy [10]. Partitional clustering is another group algorithms that require a user to input a target number of clusters k so that the clustering process merges data to k clusters. K-means is a popular partitional clustering algorithm [11]. Based on the user-specified k, the algorithm starts from randomly selecting k objects as starting centroids. Afterward, the process iteratively reassigns the objects into k disjoint groups based on the similarity measure among each centroid and its group objects. The main challenge of the K-means is to accurately identify the real number of clusters. Clustering is also a popular method to identify potential groups that have different characteristics from a disease population, such as asthma [12].

A New Technique for Comprehensive Association Discovery: Association Rule Mining

Principle of Association Rule Mining

Unlike classification for classifying new data and clustering for discovering hidden structure of the dataset, association rule mining (ARM) is a method to discover meaningful relations between variables in databases. Agrawal et al. first introduced the concept of ARM to extract regularities between products in large-scale warehouse databases [13]. Association rules are in the form of $X \Rightarrow Y$, which means that X implies Y, where X and Y are called antecedent and consequent, respectively [14]. For example, in a cancer study, a rule such as $\{TumorSize = Large\ and\ Irradiation = No\} \Rightarrow \{Recurrence = High\}$ implies that "if a patient has a large-sized tumor scanned and no record of irradiation treatment, the patient may have a high chance of cancer recurrence."

Support and confidence are two important metrics quantifying a rule's frequency and the level of association. The support of an association rule is defined as the fraction of the records in the database that contain both X and Y. A high support of an association indicates that a high portion of the database is applicable to the rule (i.e., frequent). The confidence of an association rule measures the ratio of records that contain all items in both X and Y to the records that only contain items in X, which reveals the level of association. The mining process requires users to specify a minimum support ($Supp_{min}$) and a minimum confidence ($Conf_{min}$) to drop infrequent and unconfident rules. The ARM process can be divided into six steps, which is depicted in Fig. 1. We provide a pseudocode of rule generation of frequent itemsets and confident rules. Interested users can refer to [15] for more detail. Improving the efficiency of mining of frequent itemset is a main research area in

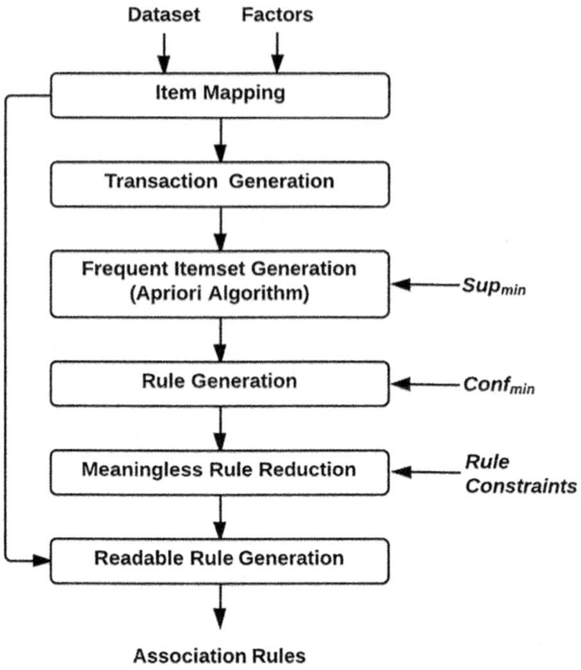

Fig. 1 Steps of association rule mining in neuropsychological dataset

ARM. One of the improvements of Apriori algorithm is called FP growth, which is based on a tree-based (called an FP-tree) representation of the given database of data tuples to considerably save amounts of memory for storing the data [16].

Pseudo-code 1: Generation of frequent itemsets

Algorithm: $F = Apriori(T, I, Supp_{min})$
// **Input:** T (Transactions), I (1-itemsets), $Supp_{min}$
// **Output:** F (Frequent Itemsets)
  $F_1 = \{f \mid f \in I, f.\text{support} \geq Supp_{min}\}$;
  for ($k = 2$; $F_{k-1} \neq \varnothing$; k++) **do**
    $C_k = GenCandidate(F_{k-1})$;
    for each transaction $t \in T$ **do**
      for each candidate $c \in C_k$ **do**
        if c is contained in t **then**
          $c.count$++;
      end
    end
  $F_k = \{c \in C_k \mid c.support \geq Supp_{min}\}$
  end
  return $F = \bigcup_k F_k$;

Pseudo-code 2: Generation of association rules

```
Algorithm: C_k = GenCandidate (F_{k-1})
// Input: F_{k-1} (Frequent k-1 itemsets)
// Output: C_k (Candidate k itemsets)
  C_k = ∅;
  forall f_m, f_n ∈ F_{k-1}
       where f_m = {i_1, … , i_{k-2}, i_{k-1}}
       and f_n = {i_1, … , i_{k-2}, i'_{k-1}}
       and i_{k-1} ≠ i'_{k-1} do
    c = {i_1, …, i_{k-1}, i'_{k-1}};
    C_k = C_k ∪ {c};
    foreach (k-1)-subset s of c do
              if (s ∉ F_{k-1}) then
                        delete c from C_k;
    end
  end
  return C_k;
```

ARM has several advantages making it suitable for healthcare data mining. First, unlike conventional statistical analyses that evaluate a null and alternative hypothesis, ARM can apply a variety of measures that determine the relationship in a comprehensive and flexible manner. Second, a rule's antecedent and consequent imply a direction of the relationship. Third, a rule's antecedent and consequent can consist of one or more factors, providing advanced knowledge of flexible factor interactions instead of monotonic relationship (e.g., logistic regression) [17]. Finally, ARM accepts user-specified inputs, which ensure the interestingness of each rule to optimize the mining results.

An Example of Association Rule Mining System for Healthcare Data Mining

A key component of a useful clinical decision support system is an interactive graphical user interface (GUI). In this section, we provide an application example of a system developed by our group. The system utilized ARM as the core with an interactive GUI for effective and real-time evidence search. The tool was implemented in MATLAB and was designed to be highly compatible with comprehensive clinical settings.

The system's GUI consists of two main windows. The first *Item Management* window (Fig. 2a) allows users to construct new items that can be used in rule antecedents and consequents. The user can specify the position of the new item to appear only in antecedent, only in consequent, or without constrain. Doing so can remove non-intuitive rules, such as $\{IQ < 70\} \Rightarrow \{Age < 11\}$ since the age cannot be the outcome of low IQ scores. The second main window is the *Rule Mining* window (Fig. 2b). It allows users to assign defined items in antecedents and consequents and generates all rules that contain these items. The user can prune out infrequent and/or unconfident rules by increasing $Supp_{min}$ and/or $Conf_{min}$,

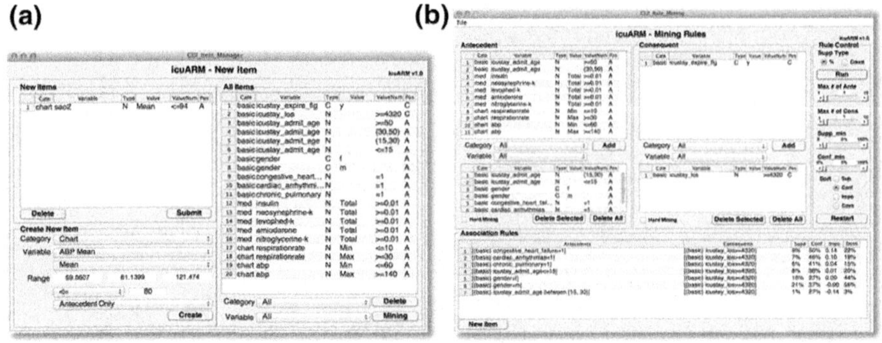

Fig. 2 **a** Rule Mining window and **b** New Item window

Table 1 Rules predicting general mental problem (SF36M < 45.91)

Rule #	Antecedent	*Supp* (%)	*Conf* (%)
1	BDI + PSSE	6.8	82.0
2	BDI + FAD	5.5	81.9

respectively. Users can also manipulate rules by specifying rule length (i.e., the number of items) or sorting rules by their supports or confidences.

An Example of the Association Rule Mining System in Predictive Health

The feasibility of our ARM system has been demonstrated previously in diverse healthcare settings, including pediatric neuropsychology [18] and intensive care units [19]. In this section we provide the third use case in the application of predictive health setting.

Predictive health (PH) is a new and innovative healthcare model that focuses on maintaining health rather than curing diseases. Computer-based decision support systems may benefit this domain by providing more quantitative health assessment, enabling more objective advice and action plans from predictive health providers. However, data mining for predictive health is more challenging compared to that for diseases. For example, because the phenotype of health relies on interactions not limited to biology, PH data also contains measurements from multiple disciplines to provide a comprehensive description of human health. However, multidisciplinary data implies information heterogeneity among measurements, which is a common challenge in healthcare data mining. This is a reason why decision support systems are rare in the domain of predictive health.

In this case study we utilize our ARM system to generate quantitative and objective rules for health assessment and prediction. This case study is conducted in conjunction with Emory Center for Health Discovery and Well Being (CHDWB®). The dataset contains 2,637 de-identified health reports from 696 healthy participants with 906 measurement variables. This case study results in 12 rules that predict mental illness based on five psychological factors, allowing us to provide important knowledge to prevent the development of mental illness. For example, Table 1 lists two rules from these 12 rules. By reading these two rules, we know that if a person has developed depressive symptoms (BDI), providers need to offer proactive advice for the prevention of the potential development of disorders especially in perceived empathic self-efficacy (PSSE) and family functioning (FAD) because they are associated with mental illness risk if comorbid with BDI. More information about this case study can be referred in [20].

Challenges of Current Association Rule Mining and Possible Solutions

Although association rule mining techniques can be elegant and powerful tools to extract meaningful patterns from healthcare data, a few remaining challenges should be highlighted. Possible solutions to address these challenges would benefit further adoption of ARM approaches by the community of healthcare informatics.

First, the first step of current ARM requires users to manually specify attributes of items in antecedents and consequents. However, not all attributes may be determined to best represent associations. Therefore, the mining results would not be objective and optimal from such user-specified attributes. This is even more challenging in high-dimensional healthcare database that consists of hundreds of attributes so that manual assignment of attributes becomes difficult. To address this challenge, during the item construction phase, users can first utilize feature selection to remove redundant or irrelevant features [21]. Given a consequent itemset as a class, supervised feature selection, such as Minimum redundancy and maximum relevance (mRMR) algorithm can be considered [22].

The second challenge lies in the manual assignment of cut-points to discretize numerical attributes into categorical attributes. Since users may not always know the optimal cut-points that can produce optimal association rules, computer-based discretization methods are needed to provide optimal cut-points so as to maximize the interestingness of rules. Researchers can consider the RUDE algorithm that provides a global discretization strategy that was originally designed for association rule mining [23]. Given a set of determinant features and optimal cut-points, users can confidently construct items and generate more reliable rules.

Third, current ARM often produces too many rules. Existing research has shown that most of the discovered rules are actually insignificant [24, 25]. Some basic pruning techniques, such as chi-square test [26], can be considered to remove those spurious or insignificant rules. In addition, being a significant (or nonredundant) rule, however, does not mean that it is a potentially actionable rule. In some domains, it is difficult to perform actions using rules with too many conditions, and/or with attributes that are hard to act upon, even though the rules are confident and significant. Such non-actionable rules should be further identified [27].

Fourth, a majority of clinical decision-making is influenced by sequential or causal relationships between events. Current ARM methods that only consider item coexistence should be expanded to sequential ARM (SARM). Sequential rules are noted by $X \Rightarrow_T Y$ to describe patterns of antecedent X followed by the consequent Y within a specific time window of length T. Methods such as MUTARA can be considered to mine unexpected temporal association rule (UTAR) from infrequent sequential patterns [28].

Finally, current ARM mainly represents rules using tables. However, the mining process often generates too many rules to be handled comfortably by humans. Clinicians may not easily find the rule that is most relevant to the patient by browsing rows line-by-line. Many association rule visualization techniques have been proposed [29–32], but all of them are designed to summarize all rules instead of searching a specific rule. Therefore, we need to provide interactive, user-friendly, and real-time visualization techniques for care providers to effectly find specific rules that can best describe a patient's conditions.

Conclusion

The healthcare industry today generates large amounts of complex data. The large amount of data is a key resource to be processed and analyzed into knowledge that enables accurate, productive, and low-cost support for decision-making. In this chapter, we have introduced the background of data mining in health care by providing several key mining methods and applications. We then moved our focus and described an evidence-based data mining method, called association rule mining, with its key advantages. Afterward, we provided an example system with interactive graphical user interface and demonstrated the system's usability using a data in Predictive Health setting.

References

1. I. Yoo, P. Alafaireet, M. Marinov, K. Pena-Hernandez, R. Gopidi, J.-F. Chang et al., Data mining in healthcare and biomedicine: a survey of the literature. J. Med. Syst. **36**, 2431–2448 (2012)
2. D. Delen, G. Walker, A. Kadam, Predicting breast cancer survivability: a comparison of three data mining methods. Artif. Intell. Med. **34**, 113–127 (2005)
3. J.A. Anderson, *An Introduction to Neural Networks* (MIT press, 1995)
4. J.R. Quinlan, Induction of decision trees. Mach. Learn. **1**, 81–106 (1986)
5. D. Meyer, F. Leisch, K. Hornik, The support vector machine under test. Neurocomputing **55**, 169–186 (2003)

6. L.-C. Huang, S.-Y. Hsu, E. Lin, A comparison of classification methods for predicting Chronic Fatigue Syndrome based on genetic data. J. Transl. Med. **7**, 81 (2009)
7. J. Yu, S. Ongarello, R. Fiedler, X. Chen, G. Toffolo, C. Cobelli et al., Ovarian cancer identification based on dimensionality reduction for high-throughput mass spectrometry data. Bioinformatics **21**, 2200–2209 (2005)
8. O. Salem, A. Guerassimov, A. Mehaoua, et al., Anomaly detection in medical wireless sensor networks using SVM and linear regression models. Int. J. E-Health. Med. Commun. **5**(1), 20–45 (2016)
9. M. Weber, M. Welling, P. Perona, *Unsupervised Learning of Models for Recognition* (Springer, 2000)
10. I. Yoo, X. Hu, A comprehensive comparison study of document clustering for a biomedical digital library MEDLINE, in *Proceedings of the 6th ACM/IEEE-CS Joint Conference on Digital libraries* (2006), pp. 220–229
11. I. Yoo, X. Hu, I.-Y. Song, Biomedical ontology improves biomedical literature clustering performance: a comparison study. Int. J. Bioinform. Res. Appl. **3**, 414–428 (2007)
12. P. Haldar, I.D. Pavord, D.E. Shaw, M.A. Berry, M. Thomas, C.E. Brightling et al., Cluster analysis and clinical asthma phenotypes. Am. J. Respir. Crit. Care Med. **178**, 218–224 (2008)
13. R. Agrawal, T. Imieliński, A. Swami, Mining association rules between sets of items in large databases, in *ACM SIGMOD Record* (1993), pp. 207–216
14. J. Hipp, U. Güntzer, G. Nakhaeizadeh, Algorithms for association rule mining—a general survey and comparison. ACM SIGKDD Explor. Newslett. **2**, 58–64 (2000)
15. R. Agrawal, R. Srikant, Fast algorithms for mining association rules, in *Proceedings of the 20th International Conference on Very Large Data Bases, VLDB* (1994), pp. 487–499
16. J. Han, J. Pei, Y. Yin, Mining frequent patterns without candidate generation, in *ACM Sigmod Record* (2000), pp. 1–12
17. P. Laxminarayan, S.A. Alvarez, C. Ruiz, M. Moonis, Mining statistically significant associations for exploratory analysis of human sleep data. IEEE Trans. Inf. Technol. Biomed. **10**, 440–450 (2006)
18. C.-W. Cheng, G.S. Martin, P.-Y. Wu, M.D. Wang, PHARM-Association Rule Mining for Predictive Health, in *The International Conference on Health Informatics* (2013), pp. 114–117
19. C. Cheng, N. Chanani, J. Venugopalan, K. Maher, D. Wang, icuARM—an ICU clinical decision support system using association rule mining. IEEE J. Trans. Eng. Health Med. **21** (1), 4400110 (2013)
20. C.-W. Cheng, G.S. Martin, P.-Y. Wu, M.D. Wang, PHARM-Association Rule Mining for Predictive Health, in *The International Conference on Health Informatics* (2014), pp. 114–117
21. Y. Saeys, I. Inza, P. Larrañaga, A review of feature selection techniques in bioinformatics. Bioinformatics **23**, 2507–2517 (2007)
22. H. Peng, C. Ding, F. Long, Minimum redundancy maximum relevance feature selection. IEEE Intell. Syst. **20**(6), 70–71 (2005)
23. M.-C. Lud, G. Widmer, Relative unsupervised discretization for association rule mining, in *Principles of Data Mining and Knowledge Discovery* (Springer, 2000), pp. 148–158
24. B. Goethals, J. Muhonen, H. Toivonen, Mining non-derivable association rules, in *SDM* (2005)
25. M.J. Zaki, Mining non-redundant association rules. Data Min. Knowl. Disc. **9**, 223–248 (2004)
26. B. Liu, W. Hsu, S. Chen, Y. Ma, Analyzing the subjective interestingness of association rules. IEEE Intell. Syst. Appl. **15**, 47–55 (2000)
27. B. Liu, W. Hsu, Y. Ma, Identifying non-actionable association rules, in *Proceedings of the Seventh ACM SIGKDD International Conference on Knowledge Discovery And Data Mining* (2001), pp. 329–334

28. H. Jin, J. Chen, H. He, G.J. Williams, C. Kelman, C.M. O'Keefe, Mining unexpected temporal associations: applications in detecting adverse drug reactions. IEEE Trans. Inf. Technol. Biomed. **12**, 488–500 (2008)
29. R.J. Bayardo Jr., R. Agrawal, Mining the most interesting rules, in *Proceedings of the Fifth ACM SIGKDD International Conference on Knowledge Discovery And Data Mining* (1999), pp. 145–154
30. A. Unwin, H. Hofmann, K. Bernt, The TwoKey plot for multiple association rules control, in *Principles of Data Mining and Knowledge Discovery* (Springer, 2001), pp. 472–483
31. K.-H. Ong, K.-L. Ong, W.-K. Ng, E.-P. Lim, Crystalclear: active visualization of association rules, in *ICDM-02 Workshop on Active Mining (AM-02)* (2002)
32. P. Buono, M.F. Costabile, Visualizing association rules in a framework for visual data mining, in *From Integrated Publication and Information Systems to Information and Knowledge Environments* (Springer, 2005), pp. 221–231

Printed by Printforce, the Netherlands